Progress in Theoretical Computer Science

Julian Charles Bradfield

Verifying Temporal Properties of Systems

Birkhäuser
Boston • Basel • Berlin

Julian Charles Bradfield
Department of Computer Science
University of Edinburgh
The King's Building
Mayfield Road
Edinburgh
The United Kingdom
EH9 3JZ

Printed on acid-free paper.

ISBN-13: 978-1-4684-6821-2   e-ISBN-13: 978-1-4684-6819-9
DOI: 10.1007/978-1-4684-6819-9

Camera ready text prepared by the Author.

Contents

Preface

This monograph aims to provide a powerful general-purpose proof technique for the verification of systems, whether finite or infinite. It extends the idea of finite local model-checking, which was introduced by Stirling and Walker: rather than traversing the entire state space of a model, as is done for model-checking in the sense of Emerson, Clarke *et al.* (checking whether a (finite) model satisfies a formula), local model-checking asks whether a particular state satisfies a formula, and only explores the nearby states far enough to answer that question. The technique used was a tableau method, constructing a tableau according to the formula and the local structure of the model. This tableau technique is here generalized to the infinite case by considering sets of states, rather than single states; because the logic used, the propositional modal mu-calculus, separates simple modal and boolean connectives from powerful fix-point operators (which make the logic more expressive than many other temporal logics), it is possible to give a relatively straightforward set of rules for constructing a tableau. Much of the subtlety is removed from the tableau itself, and put into a relation on the state space defined by the tableau—the success of the tableau then depends on the well-foundedness of this relation.

The generalized tableau technique is exhibited on Petri nets, and various standard notions from net theory are shown to play a part in the use of the technique on nets—in particular, the invariant calculus has a major role.

The requirement for a finite presentation of tableaux for infinite systems raises the question of the expressive power of the mu-calculus. This is studied in some detail, and it is shown that on reasonably powerful models of computation, such as Petri nets, the mu-calculus can express properties that are not merely undecidable, but not even arithmetical.

This monograph is based on my doctoral dissertation, examined in May 1991 at the University of Edinburgh; some of the material has been published in [Bra91], [BrS90] and [BrS91].

Acknowledgements

Most of this work was done while I was a research student in the Department of Computer Science at the University of Edinburgh, supported by an award from the U.K. Science and Engineering Research Council; the remainder was done while I was a research assistant on the SERC grant 'Mathematically Proven Safety Systems' (GR/F 38808) at Edinburgh.

I thank especially my thesis supervisor, Colin Stirling, without whose guidance, encouragement and criticism this work would not have appeared.

My thesis examiners, Eike Best and Robin Milner, suggested some improvements; Robin Milner suggested producing this monograph.

I have benefited from conversations with many people in Edinburgh, particularly Mads Dam and Glenn Bruns.

I thank Glynn Winskel, who introduced me to temporal logic and model-checking, and suggested that I go to Edinburgh.

Finally, I thank Perdita Stevens for moral support throughout.

Chapter 1

Introduction

1.1 Infinite state model-checking.

This monograph is concerned with the verification of infinite systems. 'Verification' has connotations of algorithmic checking, and is chosen for that reason, for the topic is the combination of two areas which have hitherto been considered separately.

Verification in its widest sense has been a major research topic since the beginnings of computer science. For some time, effort was directed towards proving properties of programs by means of logical reasoning, with the meaning of programs given either by logic also—whether predicate logic, as with Floyd, de Bakker, Park *et al.*, or temporal logic, as with Manna and Pnueli—or by a denotational or operational semantics.

About ten years ago, a new approach was begun by Clarke, Emerson, Sifakis and others. This approach is termed 'model-checking', since the idea is to consider some system as a model for some logic, and check whether the model satisfies a given formula of the logic expressing some desirable property. The distinctive feature is *checking*: rather than performing proofs, one has an algorithm which takes the model and formula as input and returns a yes/no answer. Clarke *et al.* developed algorithms for the logic CTL (described later), and their algorithms have been implemented by themselves and others (including an implementation in purely functional ML by this author), and have produced useful results.

Also ten years ago, Pratt and Kozen introduced a temporal logic called the 'modal mu-calculus'. This logic combines standard modal logic with least and greatest fix-point operators to produce a remarkably expressive logic. Although it looks 'modal', in that no mention is made of paths, the fix-point operators allow the expression of very complex 'temporal' properties, that is, properties involving paths. The modal mu-calculus subsumes many other temporal logics, and so its study is especially useful. Amongst other research, the model-checking idea was transferred to the modal mu-calculus by Emerson and Lei.

However, these model-checking algorithms all proceed by an exhaustive traversal of the state space of the model. Therefore, they are inherently incapable of considering *infinite* systems. Moreover, this exhaustive traversal may well be unnecessary—some properties depend only on very small parts of a system.

On the other hand, infinite systems and potentially infinite systems, which become more common as interest develops in concurrent and distributed systems, are of course amenable to the logical attack mentioned first. The problem here is that even when a complete proof system is available, an effective proof system is usually not. The quest for effective proof techniques has produced much interesting research (particularly on Petri nets), but it is a fact of life that there are no general effective techniques for any but a very small class of problems.

My purpose here is to bring together the ideas of finite, algorithmic, model-checking and the ability to perform proofs. This is made easier by using the propositional modal mu-calculus, since its connectives fall into two classes. The first class comprises the modal and boolean operators, which are very simple in nature and can be 'checked'. The second comprises the fix-point operators: these, especially the least fix-point, introduce complexity which may require subtle techniques to analyse.

1.2 Background.

1.2.1 Imperative programs and Hoare logic.

The formal proof of program correctness began in the late 1960s. Floyd [Flo67] proposed a method of assigning meanings to imperative languages by annotating each point in the control flow with a proposition, in some logic such as predicate calculus, which should hold there, and he gave techniques for doing this. Floyd's work was formalized and developed by Manna, both alone and with Pnueli [MaP69], to turn program properties into questions of satisfiability or validity in first order logic. This was extended by Park [Par70] who considered second order logic and the use of fix-point induction to prove properties. This work considered recursive program schemes, i.e. programs with what amounts to mutually recursively defined functions. By restricting consideration to straightforward programs with loops, but no functions or procedures, Hoare [Hoa69] produced his celebrated system of axioms for partial correctness. The language studied by Hoare was an abstraction of the common contemporary languages such as Fortran and Algol, and in its simplest form is thus:

Definition 1.1. A *simple* **while**-*language* is the language of commands c built up from a class *Act* of atomic actions a and a class *Bexp* of boolean expressions b according to the following BNF rules:

$$c ::= a$$
$$\mid c_1 \; ; \; c_2$$
$$\mid \textbf{if } b \textbf{ then } c_1 \textbf{ else } c_2$$
$$\mid \textbf{while } b \textbf{ do } c$$

◁

Such programs are executed in the context of a global state that may be changed by the execution of atomic actions. The meanings of the constructors are as usually intended; the execution context also provides a means of evaluating boolean expressions (subject to the constraint of not changing the state), and then the execution of a **while** command is given by executing c repeatedly as long as b holds. The classic **while**-language has as given a class *Var* of variable names and a class *Aexp* of arithmetic expressions e, built by arithmetic operators from integer constants and variable names x; *Bexp* is built by logical operators from (in)equalities on arithmetic expressions. The global state is a memory M that is a function from *Var* to $\mathbb{Z}$; expressions are evaluated with respect to the current state, with each variable x evaluating to the value $M(x)$. The atomic actions are assignments of the form $x := e$; the execution of $x := e$ in state M results in a state M' which is the same as M except that $M'(x) = v$ where v is the result of evaluating e in state M. A null command **skip** which does nothing is often added.

Hoare's approach was to consider partial correctness of such programs, that is, if a program terminates then it terminates with the correct answer. He formulated partial correctness assertions of the form $\{P\}c\{Q\}$, where c is a command, and P and Q are expressions in a language of assertions about the state—usually that of arithmetic expressions on the values of variables. Such an assertion is true if whenever c is executed in a state satisfying P, the resulting state (if there is one) satisfies Q. Hoare's achievement was to provide a set of proof rules for partial correctness assertions according to the structure of c; for example, the rule for assignment is

$$\{Q[e/x]\}x := e\{Q\}$$

where $Q[e/x]$ is the result of substituting e for each free occurrence of x in Q; and the rule for **while** statements is

$$\frac{\{I \wedge b\}c\{I\}}{\{I\}\textbf{while } b \textbf{ do } c\{I \wedge \neg b\}}$$

This rule uses the concept of invariant—since we do not know how many times c will be executed, all we can expect to prove is that some property remains unchanged. Now an invariant is a fix-point, so this is an early appearance of fix-points in a formal proof system for program properties.

Meanwhile, the classical approach to the denotational semantics of programs was being developed by Scott and Strachey ([Sco70] and [ScS71]), using complete partial orders; here fix-points appeared explicitly, in giving the denotation of **while** commands. Using this semantics, it was not difficult to show that Hoare's rules were sound; moreover, in [Coo78] it was shown that a relative completeness result holds: the Hoare rules are complete provided that (i) one can decide the assertion language (whether by means of a complete proof system, or by an oracle) and (ii) the assertion language is strong enough to express the weakest precondition P of a program c with respect to a postcondition Q (i.e. $\forall P' . \{P'\}c\{Q\} \Rightarrow (P' \Rightarrow P)$). Questions of expressibility are often intricate; in later chapters, we shall study the expressibility of the mu-calculus, and also express Hoare logic in terms of the tableau proof system developed in chapter 3—completeness of the translation will then correspond immediately to the expressibility requirement.

Hoare logic was developed in various ways. One major problem was (and is) the extension to total correctness, that is, the ability to prove assertions $[P]c[Q]$ that not only does c terminate successfully if it terminates, but also that it does in fact terminate. Floyd [Flo67] proposed a method of well-founded sets for proving total correctness, which essentially involved, for a **while** loop, finding a well-ordering on the set of program states with respect to which the state decreased each time round the loop; as we shall see later, this appears in exactly the same form in the translation of Hoare logic.

Another direction was the extension to larger languages—in particular, languages with recursive procedures. The Scott–Strachey semantics interpreted recursive procedures as the least fix-points of functions over the semantic domain, and so it was natural to express recursion by programs of the form $\mu X.c$ (where the command c may contain occurrences of the 'program variable' X), interpreted as the least fix-point of c as a function in X, so replacing the recursive equations of program schemes by explicit fix-point constructors. By similarly extending the assertion language to allow recursive assertions $\mu Z.P$, the methods of basic Hoare logic extend to recursive procedural languages (see [Par70], [BaR72]; and [Bak80] for the full development of Hoare logic).

1.2.2 Dynamic logic.

During the 1970s, the theory of program correctness was extended by investigating more powerful logics, and studying them in a manner more similar to the traditions of mathematical logic. A family of logics which received much attention was that of dynamic logics, which can be seen as extending the ideas of Hoare logic [Pra76]. Dynamic logics are modal logics, where the different modalities correspond to the execution of different programs—the formula $\langle\alpha\rangle\Phi$ is read as 'it is possible for α to execute and result in a state satisfying Φ'. The programs may be of any type of interest; the variety of dynamic logic most often referred to is a propositional language in which the programs are built from atomic programs by regular expression constructors; henceforth, Propositional Dynamic Logic, PDL, refers to this logic, which we now define.

Definition 1.2. *Propositional Dynamic Logic* is the logic whose formulae are constructed from a set Atom of atomic propositions $P, Q, \ldots$ and a set Act of atomic programs a thus:

(i) a program α has the form

$$\alpha ::= a \mid \alpha_1; \alpha_2 \mid \alpha_1 \cup \alpha_2 \mid \alpha_1^*$$

(ii) a formula Φ has the form

$$\Phi ::= P \mid \langle\alpha\rangle\Phi_1 \mid \neg\Phi_1 \mid \Phi_1 \wedge \Phi_2$$

◁

Notation 1.3. The usual logical abbreviations are used: the boolean disjunction, $\Phi_1 \vee \Phi_2 \stackrel{\text{def}}{=} \neg(\neg\Phi_1 \wedge \neg\Phi_2)$, and implication, $\Phi_1 \Rightarrow \Phi_2 \stackrel{\text{def}}{=} \Phi_2 \vee \neg\Phi_1$, and the modal box, $[\alpha]\Phi \stackrel{\text{def}}{=} \neg\langle\alpha\rangle\neg\Phi$; henceforth, these abbreviations will be used for all relevant logics without further comment. ◁

PDL is interpreted with respect to a Kripke structure model, formalizing the notion of the global state in which programs execute and which they change—each point in the structure corresponds to a possible state, and programs determine a relation between states giving the changes effected by the programs. The semantics is defined formally by induction in the usual way, as will be given later; intuitively, $\langle \alpha^* \rangle \Phi$ means 'those states at which is possible for Φ to hold after some number of iterations of α', and $[\alpha_1 \cup \alpha_2]\Phi$ means 'after the execution of either α_1 or α_2, chosen non-deterministically, Φ will hold'.

A landmark in the study of PDL is a paper by Fischer and Ladner [FiL79] in which they showed that the satisfiability problem (i.e. does a given formula have any model?) was decidable in no more than co-NEXPTIME (and no less than DTIME($c^{n/\log n}$)). Furthermore, PDL has a small model theorem: every formula that has a model, has a finite model.

A variety of extensions and restrictions of PDL have been studied; two important extensions are test programs and looping constructors. PDL with tests extends the class of programs by a construct Φ? (so that now programs and formulae are mutually recursive), which is interpreted as a program that tests Φ; if Φ is true the program continues, and if false, the program stops (as in Dijkstra's guarded commands). PDL with tests is expressive—in particular, the programs include non-deterministic **while**-programs (**while** b **do** c translates to $(b?; c)^*; (\neg b)?$)—yet many properties of PDL are still true (in fact, the Fischer–Ladner results mentioned above were for PDL with tests). PDL-Δ, defined by Streett [Str81] adds an infinite loop operator: $\Delta\alpha$ is a formula true of states which can repeat the program α forever. PDL-Δ remains decidable, but if the program language is extended to context-free grammars it becomes (very) undecidable—the boundary was studied by Harel, Pnueli and Stavi [HPS81], who showed that the addition of the construct $\alpha^\Delta; \beta; \gamma^\Delta \overset{\text{def}}{=} \bigcup_i \alpha^i; \beta; \gamma^i$ makes the language undecidable (and indeed Π_1^1).

On the other hand, restricting the language gave worthwhile reductions in the complexity: for deterministic models (i.e. those where atomic programs are deterministic) satisfiability is decidable in exponential time [BHP82], and when programs are restricted to be (the regular expression translations of) deterministic **while**-programs, satisfiability becomes PSPACE-complete [HaR81].

Meanwhile, axiomatizations of PDL were also successfully investigated: the standard axiomatization was given in [KoP81] by Kozen and Parikh, starting from Segerberg's [Seg68] axiomatization of the modal logic S4.1, and is sound and complete.

1.2.3 Modal and temporal logic.

Mention of S4.1 leads us to the modal and temporal logic approach. The work on PDL that we have just discussed concentrated on the relationship between PDL and its models; a more 'proof-theoretic' style was developed by Manna, Pnueli and others, using more expressive logics.

Modal and temporal logics as they are used in computer science are concerned with properties of systems that have a number of states amongst which there is a relation of evolution or succession. The standard structure used in giving semantics of such logics is the transition system or Kripke structure.

Definition 1.4. A *labelled transition system* $\mathcal{T} = (\mathcal{S}, \{ \xrightarrow{a} | a \in \mathcal{L} \})$ of sort $\mathcal{L}$ is a set $\mathcal{S}$ of states, a set $\mathcal{L}$ of labels, and for each label a, a binary relation $\xrightarrow{a}$ on $\mathcal{S}$. ◁

Notation 1.5. Henceforth, any labelled transition system called $\mathcal{T}$ has the form $(\mathcal{S}, \{ \xrightarrow{a} | a \in \mathcal{L} \})$ unless otherwise stated. Similarly for $\mathcal{T}'$ etc.

If $K \subseteq \mathcal{L}$, the notation $s \xrightarrow{K} s'$ means $\exists a \in K . s \xrightarrow{a} s'$. ◁

Unlabelled transition systems, or Kripke structures, are considered to be labelled transition systems with some singleton label set.

Very many systems can be viewed as labelled transition systems: **while**-programs, where each possible memory configuration is a state, and the labels are programs; the PDL structures defined earlier; Petri nets, taking markings as states and transitions (or (multi)sets of transitions in a non-interleaving approach) as labels; CCS processes, taking processes as states and actions as labels; and so on.

With respect to a transition system model (and if appropriate a valuation of atomic propositions) modal connectives such as 'for all successor states' can be interpreted in the natural way; and by considering paths through the structures, more powerful temporal properties such as 'eventually' and 'until' can be interpreted. There is a great variety of temporal logics and notations, so we now define a logic in which all the logics we mention may be expressed. The notation here is based on that of [Sti91], to which the reader is referred for information on temporal logics.

Definition 1.6. *(Full) Propositional Temporal Logic* (PTL) with a sort $\mathcal{L}$ of labels a and a set Atom of atomic propositions $P, Q, \ldots$ has the following syntax:

$$\Phi ::= P \mid \neg \Phi_1 \mid \Phi_1 \wedge \Phi_2 \mid \forall \Phi_1 \mid (K)\Phi_1 \mid \Phi_1 \, \mathbf{U} \, \Phi_2$$

where $K \subseteq \mathcal{L}$. In addition to the usual boolean abbreviations, the following abbreviations are employed: $\mathbf{F}\Phi \overset{\text{def}}{=} \text{tt } \mathbf{U} \Phi$ and $\mathbf{G}\Phi \overset{\text{def}}{=} \neg\mathbf{F}\neg\Phi$, where tt is an atomic proposition which is true everywhere, and $\bigcirc$ for $(\mathcal{L})$.   ◁

PTL formulae are properties of paths through transition systems; the formal semantics again appears in chapter 2. Intuitively, the meanings of the connectives are that $\bigcirc$ is 'next' (i.e. at the next point on the path), (K) is a relativized 'next' (i.e. 'next' via some label $a \in K$), $\mathbf{U}$ is 'until', $\forall$ is a branching connective 'for all (paths starting from the beginning of this path)', $\mathbf{F}$ is 'eventually', and $\mathbf{G}$ is 'always'.

A sublogic of PTL that does not have the branching operator $\forall$ is called a linear temporal logic.

The use of linear temporal logics such as the above in specifying and proving properties of programs was advocated by Pnueli in [Pnu81]. He argued that contemporary techniques were inadequate since they were not capable of dealing with notions of fairness, for example the question of whether a process which is infinitely often ready to run will in fact execute infinitely often, whereas such properties are expressible in linear temporal logic. Pnueli's technique was to model a concurrent program P as a set of disjoint processes executing in an interleaving fashion with fair scheduling guaranteed; he then constructed a temporal logic formula $W(P)$ describing the execution sequences of the program. Thus, to find out whether the program satisfies Φ, one tries to prove or refute $W(P) \Rightarrow \Phi$, using an axiomatization of the logic.

This approach was generalized by Manna and Pnueli [MaP83], who presented a system divided into three parts: a purely temporal logical part, comprising any standard axiomatization; a domain part, comprising axioms for specific program constructors; and a program part giving high-level axiom schemas for fairness etc. This at once gave a method applicable to many different languages and did away with much of the unreadability of Pnueli's method.

Further extensions include those of Wolper [Wol83] who (still within linear time) increased the expressive power by adding operators definable by a right linear grammar, and showed that this did not increase the complexity of the decision procedure; and Barringer, Kuiper and Pnueli [BKP84], who addressed the following issue: unlike the studies of PDL discussed earlier, in which a logic was discussed separately from its models, the Manna–Pnueli approach includes the model under verification in the proof technique, by means of the formula $W(P)$; the logic and the model are fused. This is intuitively somewhat unattractive, and moreover $W(P)$ is a very complicated formula which does not reflect well the structure of the program. Barringer *et al.* introduced a notion of environment versus program, allowing them to build a more compositional and intuitive system.

1.2.4 Modal mu-calculus.

The temporal logic used here is the propositional modal mu-calculus. This first appeared in a paper by Kozen [Koz83], who took a modal logic with labels and added a least fix-point operator, so that $\mu Z.\Phi$ is intended to be the least fix-point of Φ considered as a functional on the state space by means of its free variable Z. (In fact, a mu-calculus was also introduced by Pratt [Pra81], aiming to link PDL and the Park/De Bakker mu-calculus of programs, but Pratt's calculus had a very different semantics, employing least roots, as in recursive function theory, rather than least fix-points.) Kozen proved various results about a restricted subset of the calculus (the restriction concerned the appearance of variables in different branches of conjunctions): he gave an EXPTIME decision procedure, a small model property, and a complete proof system. Subsequently, Kozen and Parikh [KoP83] showed decidability of satisfiability for the full calculus by reducing it to Rabin's SnS. This has a non-elementary decision procedure: in [StE89] Streett and Emerson showed that the modal mu-calculus has an elementary decision procedure and a small model property.

1.2.5 Model checking.

We now turn to the other main ingredient, model-checking. The work reviewed above was chiefly concerned with problems of satisfiability and validity, which are properties of the logic rather than of its models: one asks whether there exists a model satisfying the formula. A model satisfies a formula if there is some state in the model which satisfies the formula: model-checking, as formulated by Clarke, Emerson and others, considers this question.

The first model checkers ([ClE81], [CES86]) took a fairly direct approach. Clarke and Emerson dealt with a logic called CTL (Computation Tree Logic): this is a branching time temporal logic whose formulae are interpreted on states, by taking only the subset of (unlabelled) PTL formed by restricting the temporal operators to positions immediately governed by a branching operator: so $\exists \mathbf{G} P$ and $\forall (P \mathbf{U} Q)$ are CTL formulae, but

$\forall(P \textbf{ U } \bigcirc Q)$ is not, because of the unbranched 'next'. They gave an algorithm which, given a formula Φ, traverses a transition system, labelling the states with those subformulae of Φ that hold there. The subtlety is in checking formulae of the form $\exists\textbf{G}\Phi$ ('Φ is always true on some path from here'), which is done by analysing the strongly connected components of the subgraph on which Φ is true, using a standard algorithm. (Since their models are finite and total, $\exists\textbf{G}\Phi$ can only be true at a state s if s is part of a strongly connected component of that subgraph.) This algorithm is polynomial in the size of the formula and of the state space, and can be extended to take account of some fairness considerations (of the form 'all paths must pass infinitely often through certain states').

Emerson and Lei [EmL86] subsequently extended model-checking to a more powerful logic CTL*, which combines linear and branching operators, and showed it to be no worse in complexity than model-checking for linear time logic (PSPACE-complete [SiC86], but of low complexity in the size of the state space and high in the size of the formula).

Subsequent research into CTL model-checking has concentrated on techniques to reduce the complexity of checking large systems, for example systems composed of many identical components—see, e.g. [BCG89] for the development of an 'indexed logic' for dealing with such systems.

1.3 Local model-checking and infinite systems.

The work in this monograph grew out of that of Stirling and Walker ([StW89], [StW90]), who were concerned to address the global nature of work on model-checking. Stirling had advocated the use of tableau techniques, which are local: that is, the properties of a given state are checked by reference to properties of adjacent states. The questions asked are different: model-checking in the sense of Clarke *et al.* asks 'does this *model* satisfy the formula?', meaning 'does there exist a state in the model at which the formula holds?', whereas local model-checking asks directly 'does this *state* in the model satisfy the formula?'

Stirling and Walker developed a tableau system for local model-checking of finite systems, using the propositional modal mu-calculus to express properties. This system provided a very powerful model-checking algorithm, and since it is local, it was natural to wonder what would happen if one tried to apply the same idea to the infinite case. What happens is that one does indeed obtain a very powerful proof technique for infinite systems, which 'checks' the checkable components of a property in question, while still leaving room for 'proving' the complex fix-point properties. Moreover, the technique provides substantial insight into the very subtle nature of mu-calculus properties, and so is of considerable interest in its own right.

This technique requires some intelligence for its use—in general, considerable knowledge of the model or class of models may be needed. Therefore, a major concern is how the nature of models interacts with the logic so that it can be exploited when performing model-checking in this infinite-state tableau system. I have chosen Petri nets as a first class of models to study, and it turns out that some of the standard notions of net theory are important in practice when doing infinite-state model-checking on nets. Moreover, some of the questions raised in this area lead on to deep foundational questions about the modal mu-calculus, a topic which is discussed in the fifth chapter.

1.4 Synopsis.

Chapter 2 starts by providing the formal semantics etc. for the various logics mentioned in this introduction.

The primary aim of the chapter is to explain the modal mu-calculus (a slight extension of Kozen's version). I define the logic and its semantics, and consider some very simple examples. I introduce the key idea of ordinal approximants. After pausing to exhibit translations of two earlier logics into the modal mu-calculus, I move on to the difficult notion of alternating fix-point properties, introducing notations for the alternation hierarchies and analysing examples. Finally I demonstrate how the slight extension to the original calculus allows the expression of 'dynamic' properties such as events happening.

Chapter 3 is the main body of the book, in which I develop a tableau system for infinite state model-checking. I first give an informal explanation of how the system works, showing how the introduction of 'propositional constants' is used to keep track of unfolding fix-points. The formal definition of the proof trees that form tableaux follows, and then I give the delicate conditions which determine whether a tableau is successful in proving its root. This success depends on the well-foundedness of a relation defined by the tableau, a relation that can be seen as encapsulating the idea of the truth of a fix-point at one state depending on the truth at certain nearby states; this is the essential combination of local model-checking and infinite state spaces. The use of logical proof techniques appears in the problem of proving that this relation is indeed well-founded, which is not part of the tableau system proper: to prove well-foundedness, any necessary reasoning may be used.

After some simple examples, I prove the soundness of the system by an inductive argument on the height of tableaux, making essential use of the notion of ordinal approximants. Rather easier is the completeness proof, which uses a notion of 'signature' to guide the (non-effective) construction of a successful tableau.

I end the chapter by considering variants of the system and derived rules, and by giving a detailed explanation of how Hoare logic can be seen in tableau terms.

Chapter 4 considers the application of the system developed in chapter 3 to concurrent systems in the form of Petri nets. I start with a brief introduction to nets, selecting a few of the many notions and results about nets that may be of use in constructing tableaux. I then consider some of the ways in which nets may be combined, again choosing some approaches which I believe will be important in future development of this work.

After this introductory material, I describe in detail the application of the tableau system to prove some safety and liveness properties of nets, intending to provide enough complexity to exercise the features of the system while still maintaining easy understanding of the examples. In these examples emerges the importance of the net-theoretic notion of invariant in applications of the tableau system to nets.

Next I discuss a technique for dealing with replicated systems. The idea is simple, namely to parametrize tableaux. However, unlike approaches such as indexed CTL [BCG89], I do not incorporate indexing or parametrization into the logic, but rather keep it in the meta-language, so that one aims to produce a proof schema for a schematic tableau. I exhibit this technique on the classic resource-control problem, showing again the importance of invariants.

I move on to consider the application of a different technique from net theory, the coverability graph. This is a device invented to answer certain questions about liveness and deadlock, and I show how it can be a useful guide to the construction of tableaux.

Lastly in this chapter, I make some remarks on compositionality. The exploitation of compositionality is a difficult topic which I intend to pursue in later work; here I just show how the behaviour of net invariants under certain constructions may help in the difficult initial stage of finding a suitable set from which to start the construction of a tableau.

Chapter 5 discusses the complexity of mu-formulae on nets. This can be seen as the study of local expressibility, i.e. expressibility *within* a model, to go with local model-checking, whereas traditional questions of expressibility *about* models sit naturally with global model-checking. Since to build a tableau one must be able to write down the sets of states involved, this matter has practical importance; but it is also of great intrinsic interest. It turns out that the modal mu-calculus can express sets of states which are not only not decidable, but not even arithmetical. The techniques used here lead to intriguing connexions between the modal mu-calculus and traditional (and less traditional) definability problems, which will be a fascinating area for future work.

I close by presenting my conclusions and intentions for further work.

Chapter 2

Program Logics and the Mu-Calculus

In this chapter, I give the formal semantics of the temporal logics mentioned earlier; then I define and exemplify the modal mu-calculus.

2.1 Semantics of temporal logics.

We consider first Propositional Dynamic Logic. A model for PDL is just a set of states and a way of interpreting atomic programs as transitions between states:

Definition 2.1. A PDL *model* is a set $\mathcal{W}$ together with a valuation $\mathcal{V} = (\mathcal{V}_{\text{Atom}}: \text{Atom} \to 2^{\mathcal{W}}, \mathcal{V}_{\text{Act}}: \text{Act} \to 2^{\mathcal{W} \times \mathcal{W}})$.

The *denotation* $\|\alpha\|_{\mathcal{V}}^{\mathcal{W}}$ of a program α in the model $(\mathcal{W}, \mathcal{V})$ is a binary relation on $\mathcal{W}$ given by the following rules, where as usual $\circ$ is relation composition and $*$ is reflexive and transitive closure:

$$\|a\|_{\mathcal{V}}^{\mathcal{W}} = \mathcal{V}_{\text{Act}}(a)$$

$$\|\alpha_1; \alpha_2\|_{\mathcal{V}}^{\mathcal{W}} = \|\alpha_1\|_{\mathcal{V}}^{\mathcal{W}} \circ \|\alpha_2\|_{\mathcal{V}}^{\mathcal{W}}$$

$$\|\alpha_1 \cup \alpha_2\|_{\mathcal{V}}^{\mathcal{W}} = \|\alpha_1\|_{\mathcal{V}}^{\mathcal{W}} \cup \|\alpha_2\|_{\mathcal{V}}^{\mathcal{W}}$$

$$\|\alpha^*\|_{\mathcal{V}}^{\mathcal{W}} = (\|\alpha\|_{\mathcal{V}}^{\mathcal{W}})^*$$

The denotation $\|\Phi\|_{\mathcal{V}}^{\mathcal{W}}$ of a formula is a subset of $\mathcal{W}$, given by

$$\|P\|_{\mathcal{V}}^{\mathcal{W}} = \mathcal{V}_{\text{Atom}}(P)$$

$$\|\neg\Phi\|_{\mathcal{V}}^{\mathcal{W}} = \mathcal{W} - \|\Phi\|_{\mathcal{V}}^{\mathcal{W}}$$

$$\|\Phi_1 \wedge \Phi_2\|_{\mathcal{V}}^{\mathcal{W}} = \|\Phi_1\|_{\mathcal{V}}^{\mathcal{W}} \cap \|\Phi_2\|_{\mathcal{V}}^{\mathcal{W}}$$

$$\|\langle\alpha\rangle\Phi\|_{\mathcal{V}}^{\mathcal{W}} = \{\, w \in \mathcal{W} \mid \exists w' \in \mathcal{W} . (w, w') \in \|\alpha\|_{\mathcal{V}}^{\mathcal{W}} \wedge w' \in \|\Phi\|_{\mathcal{V}}^{\mathcal{W}} \,\}$$

◁

Notation 2.2. Where they can be understood from context, the indices $\mathcal{W}$ and $\mathcal{V}$ will be omitted from the denotational brackets $\| \cdot \|_{\mathcal{V}}^{\mathcal{W}}$.

$W \vDash \Phi$ means $W \subseteq \|\Phi\|$. ◁

The various extensions to PDL are defined similarly: the meaning of the test operator is

$$\|\Phi?\|_{\mathcal{V}}^{\mathcal{W}} = \{ (w, w) \in \mathcal{W} \times \mathcal{W} \mid w \in \|\Phi\|_{\mathcal{V}}^{\mathcal{W}} \}.$$

and the meaning of the loop operator is given by an explicit fix-point construction:

$$\|\Delta\alpha\|_{\mathcal{V}}^{\mathcal{W}} = \bigcup \{ X \subseteq W \mid w \in X \Rightarrow \exists w' . (w, w') \in \|\alpha\|_{\mathcal{V}}^{\mathcal{W}} \wedge w' \in X \}$$

Propositional Temporal Logic, on the other hand, takes maximal paths through the system as its primary objects, rather than states:

Definition 2.3. In a transition system $\mathcal{T}$ a *path* π is a (finite or infinite) sequence $s_0 \xrightarrow{a_0} s_1 \xrightarrow{a_1} \cdots$ of states s_i and labels a_i. $\pi(i)$ denotes s_i, the ith state; $\mathcal{L}(\pi, i)$ is a_i, the ith label, and π^i is the path $s_i \xrightarrow{a_i} \cdots$, the ith suffix. π is a *run* if it is maximal, i.e. either it is infinite or it is finite of length n and there is no a', s' such that $s_n \xrightarrow{a'} s'$. ◁

The meaning of a PTL formula in a given transition system is then a set of runs:

Definition 2.4. A PTL *model* is a transition system $\mathcal{T}$ of sort $\mathcal{L}$, together with a valuation $\mathcal{V} \colon \text{Atom} \to 2^S$. In such a model the denotation of a PTL formula is a set of runs through the system, according to the following rules:

$$\|P\|_{\mathcal{V}}^{\mathcal{T}} = \{ \pi \in \mathcal{R} \mid \pi(0) \in \mathcal{V}(P) \}$$

$$\|\neg\Phi\|_{\mathcal{V}}^{\mathcal{T}} = \mathcal{R} - \|\Phi\|_{\mathcal{V}}^{\mathcal{T}}$$

$$\|\Phi_1 \wedge \Phi_2\|_{\mathcal{V}}^{\mathcal{T}} = \|\Phi_1\|_{\mathcal{V}}^{\mathcal{T}} \cap \|\Phi_2\|_{\mathcal{V}}^{\mathcal{T}}$$

$$\|\forall\Phi\|_{\mathcal{V}}^{\mathcal{T}} = \{ \pi \in \mathcal{R} \mid \forall\pi' \in \mathcal{R} . \pi'(0) = \pi(0) \Rightarrow \pi' \in \|\Phi\|_{\mathcal{V}}^{\mathcal{T}} \}$$

$$\|(K)\Phi\|_{\mathcal{V}}^{\mathcal{T}} = \{ \pi \in \mathcal{R} \mid \mathcal{L}(\pi, 0) \in K \wedge \pi^1 \in \|\Phi\|_{\mathcal{V}}^{\mathcal{T}} \}$$

$$\|\Phi_1 \, \mathbf{U} \, \Phi_2\|_{\mathcal{V}}^{\mathcal{T}} = \{ \pi \in \mathcal{R} \mid \exists i . \pi^i \in \|\Phi_2\|_{\mathcal{V}}^{\mathcal{T}} \wedge \forall j < i . \pi^j \in \|\Phi_1\|_{\mathcal{V}}^{\mathcal{T}} \}$$

where $\mathcal{R}$ is the set of all runs. ◁

Sometimes the set of runs is restricted to some given set of 'fair' runs; this is a means of incorporating fairness without expressing it in the logic. We do not consider this.

2.2 The propositional modal mu-calculus.

The remainder of this chapter deals with modal mu-calculus. In this book we use a slight extension of Kozen's mu-calculus, in that we use modalities indexed by sets of labels rather than by single labels—this allows the concise expression of useful properties, as demonstrated in the examples. Without further ado:

Definition 2.5. The *propositional modal mu-calculus* has formulae Φ built from a set Var of variables $X, Y, Z, \ldots$ and a set $\mathcal{L}$ of labels $a, b, \ldots$ by the following rules:

$$\Phi ::= Z \mid \neg\Phi_1 \mid \Phi_1 \wedge \Phi_2 \mid [K]\Phi_1 \mid \nu Z.\Phi_1$$

where K ranges over subsets of $\mathcal{L}$, and $\nu Z.\Phi_1$ is subject to the restriction that any free occurrence of Z in Φ_1 must be within the scope of an even number of negation symbols. The usual boolean and modal dual connectives are employed ($\langle K \rangle \Phi \stackrel{\text{def}}{=} \neg[K]\neg\Phi$), and a dual to ν is defined by

$$\mu Z.\Phi \stackrel{\text{def}}{=} \neg\nu Z.\neg\Phi[\neg Z/Z]$$

where $\Phi[\Psi/Z]$ denotes the syntactic substitution of Ψ for free occurrences of Z in Φ. The boolean constants are defined by abbreviation: $\text{tt} \stackrel{\text{def}}{=} \nu Z.Z$ and $\text{ff} \stackrel{\text{def}}{=} \neg\text{tt}$.

Further, the following abbreviations are used:

$$[-K]\Phi \stackrel{\text{def}}{=} [\mathcal{L} - K]\Phi$$

$$[a_1, \ldots, a_n]\Phi \stackrel{\text{def}}{=} [\{a_1, \ldots, a_n\}]\Phi$$

$$[-]\Phi \stackrel{\text{def}}{=} [\mathcal{L}]\Phi$$

and similarly for $\langle K \rangle$.   ◁

Notation 2.6. $P, Q, \ldots$ will be used to range over a subset of variable symbols which are intended to play the role of atomic propositions, so by convention they will never be bound by μ, and further any model has an intended meaning for these symbols; for example, in nets the 'atomic propositions' may be linear inequalities containing symbols for places; such formulae will only be interpreted in a net which has such places.   ◁

Notation 2.7. The word 'mu-formula' is used to mean 'formula of the mu-calculus'; but the word 'μ-formula' means a formula of the mu-calculus whose outermost connective is μ. The letter σ is used to range over μ and ν when referring to fix-points of either type.   ◁

In many results, it is convenient to think in terms of the derived operators, and forbid the explicit use of negation save when applied directly to (free) variables.

Definition 2.8. A formula of the mu-calculus extended by the dual operators defined above is said to be in *positive form* iff all negation symbols in the formula apply directly to variable symbols. Any formula can be written in positive form by using the De Morgan dualities to push $\neg$ symbols inwards. A formula is in *positive normal form* if it is in positive form and there are no clashes of bound variables, in the usual way. Any formula can be α-converted into normal form. ◁

Henceforth positive normal form will be assumed wherever convenient.

Before attempting to explain the meanings of mu-formulae, we define the semantics formally—the interpretation of complex formulae is extremely subtle, and it is important to tie intuition firmly to the semantics—or *vice versa*!

Definition 2.9. A mu-calculus *model* is a labelled transition system T of sort $\mathcal{L}$ together with a valuation $\mathcal{V} \colon \mathrm{Var} \to 2^S$. The denotation $\|\Phi\|_{\mathcal{V}}^T$ of a mu-formula Φ in the model $(T, \mathcal{V})$ is given by the following rules (omitting the superscript T):

$$\|Z\|_{\mathcal{V}} = \mathcal{V}(Z)$$

$$\|\neg\Phi\|_{\mathcal{V}} = S - \|\Phi\|_{\mathcal{V}}$$

$$\|\Phi_1 \wedge \Phi_2\|_{\mathcal{V}} = \|\Phi_1\|_{\mathcal{V}} \cap \|\Phi_2\|_{\mathcal{V}}$$

$$\|[K]\Phi\|_{\mathcal{V}} = \{\, s \in S \mid \forall s' \in S . \forall a \in K . s \xrightarrow{a} s' \Rightarrow s' \in \|\Phi\|_{\mathcal{V}} \,\}$$

$$\|\nu Z.\Phi\|_{\mathcal{V}} = \bigcup \{\, S \subseteq \mathcal{S} \mid \|\Phi\|_{\mathcal{V}[Z:=S]} \supseteq S \,\}$$

where $\mathcal{V}[Z := S]$ is the valuation $\mathcal{V}'$ which agrees with $\mathcal{V}$ save that $\mathcal{V}'(Z) = S$. ◁

The derived rules for the dual operators are then straightforward:

$$\|\Phi_1 \vee \Phi_2\|_V = \|\Phi_1\|_V \cup \|\Phi_2\|_V$$

$$\|\langle K \rangle \Phi\|_V = \{\, s \in \mathcal{S} \mid \exists a \in K \,.\, \exists s' \in \mathcal{S} \,.\, s \xrightarrow{a} s' \wedge s' \in \|\Phi\|_V \,\}$$

$$\|\mu Z.\Phi\|_V = \bigcap \{\, S \subseteq \mathcal{S} \mid \|\Phi\|_{V[Z := S]} \subseteq S \,\}$$

The immense expressive power of the mu-calculus comes chiefly from the ability to mix maximal and minimal fix-points. This is analogous to the alternation of existential and universal quantifiers in predicate logic (an analogy which will be developed in chapter 5).

Let us now consider simple examples of mu-formulae. At the simplest level, where there are no fixed points, we have just modal logic. Formulae with one fix-point are easy to understand, and bring out clearly the difference between minimal and maximal fix-points. The simplest such formulae are those expressing the notions of 'always' and 'eventually'. Consider any model with some atomic proposition P, and the formulae

$$\Phi \overset{\text{def}}{=} \mu Z.P \vee ([-]Z \wedge \langle - \rangle \text{tt})$$

$$\Psi \overset{\text{def}}{=} \nu Z.P \wedge [-]Z.$$

The first formula means 'eventually P'—either P holds now, or at all successor states (and there is at least one) either P holds now or ...; the second means 'always P'—P holds now, and at all successors P holds now and at all successors of these .... Why does one use a least fix-point and the other a greatest? The answer is the key to the difference between fix-points: Φ is concerned with finite behaviour, in that it requires that P eventually be true on any path in some (unbounded but) finite time, whereas Ψ is concerned with infinite behaviour, in that it requires P to be true for ever, even along infinite paths.

The distinction between finite and infinite behaviour can be understood in terms of ordinal approximants. This notion is essential to many mu-calculus results (in particular to the proof system of the next chapter), and can also be valuable in understanding mu-formulae, so we shall now spend some time on definitions and examples.

Definition 2.10. [Koz83] The syntax of the mu-calculus is extended by the addition of approximant formulae of the form $\sigma^\alpha Z.\Phi$ where α is an ordinal and Φ is an (extended) mu-formula. In addition, we sometimes use $\sigma^\infty Z.\Phi$ to mean $\sigma Z.\Phi$; ∞ is to be thought of as larger than any ordinal. ◁

Incrementing α corresponds to 'unfolding' $\sigma Z.\Phi$ in the following sense:

Definition 2.11. The *unfolding* of a σ-formula $\sigma Z.\Phi$ is the formula $\Phi[\sigma Z.\Phi/Z]$. ◁

The zeroth approximant to a least fix-point is the empty set of states, and to a greatest fix-point is the entire state space; incrementing α unfolds the formula once more, and at limit ordinals the union (intersection) of all earlier approximants is taken. Formally,

Definition 2.12. In a model (T, V), the denotation of approximant formulae is defined by the following transfinite induction:

$$\|\mu^0 Z.\Phi\|_V = \varnothing \qquad \|\nu^0 Z.\Phi\|_V = S$$

$$\|\sigma^{\alpha+1} Z.\Phi\|_V = \|\Phi\|_{V[Z:=\|\sigma^\alpha Z.\Phi\|_V]}$$

$$\|\mu^\lambda Z.\Phi\|_V = \bigcup_{\alpha<\lambda} \|\mu^\alpha Z.\Phi\|_V \qquad \|\nu^\lambda Z.\Phi\|_V = \bigcap_{\alpha<\lambda} \|\nu^\alpha Z.\Phi\|_V$$

where λ is a limit ordinal. ◁

Proposition 2.13. [Koz83] For a formula Φ in a model (T, V),

$$\|\mu Z.\Phi\|_V^T = \bigcup_{\alpha \in \mathrm{Ord}} \|\mu^\alpha Z.\Phi\|_V^T$$

and dually

$$\|\nu Z.\Phi\|_V^T = \bigcap_{\alpha \in \mathrm{Ord}} \|\nu^\alpha Z.\Phi\|_V^T$$

where Ord is the class of all ordinals. □

This proposition is the Knaster–Tarski fix-point theorem applied to the complete lattice 2^S; the theorem applies because of the

Lemma 2.14. If $\sigma Z.\Phi$, then for any model (T, V) the endofunction on 2^S given by $S \mapsto \|\Phi\|_{V[Z:=S]}$ is monotonic. □

which follows from the syntactic restriction imposed on the bodies of fix-point formulae. (This proposition explains the notation $\sigma^\infty Z.\Phi$.)

Notation 2.15. If Φ is a σ-formula $\sigma Z.\Psi$ we write Φ^α for $\sigma^\alpha Z.\Psi$. ◁

It follows from Proposition 2.13 that for a least fix-point formula Φ, a state s is in $\|\Phi\|$ iff s is in $\|\Phi^\alpha\|$ for some α; and from the definition, whether s is in $\|\Phi^\alpha\|$ depends only on $\|\Phi^\beta\|$ for $\beta < \alpha$, and since the ordinals are well-founded, the chain of dependencies eventually terminates at Φ^0; on the other hand, for a greatest fix-point formula Ψ, s is in $\|\Psi\|$ only if it is in $\|\Psi^\alpha\|$ for all α, so there is no such well-foundedness to the dependencies. Thus, the truth of least fix-points depends on something that happens in finite time on any given path, whereas greatest fix-points may depend on infinite behaviour.

For simple formulae, approximants are easily calculated and understood: if $\Phi \stackrel{\text{def}}{=} \mu Z.P \vee [-]Z$ then Φ^0 is those states where P holds (or there is no successor), Φ^1 adds states one step away from P, and so on. 'And so on' may in general extend up the ordinals, but for unnested formulae on finite-branching systems, it goes only up to ω.

Definition 2.16. For a formula $\Phi = \sigma Z.\Psi$ and a model $(\mathcal{T},\mathcal{V})$, the *closure ordinal* cl Φ of Φ with respect to $(\mathcal{T},\mathcal{V})$ is the least ordinal α such that $\|\Phi^{\alpha+1}\|_\mathcal{V}^\mathcal{T} = \|\Phi^\alpha\|_\mathcal{V}^\mathcal{T}$. ◁

Proposition 2.17.
 (i) $\|\Phi\|_\mathcal{V}^\mathcal{T} = \|\Phi^{\text{cl }\Phi}\|_\mathcal{V}^\mathcal{T}$ (immediate from definitions)
 (ii) $|\text{cl }\Phi| \leq |\mathcal{S}|$ (by a simple cardinality argument). □

Proposition 2.18. [Lar90] If $(\mathcal{T},\mathcal{V})$ is a finite-branching system (that is, $\forall a \in \mathcal{L}. \forall s \in \mathcal{S}. |\{\, s' \mid s \stackrel{a}{\longrightarrow} s' \,\}| < \aleph_0$) and $\Phi = \sigma Z.\Psi$ is unnested (see Definition 2.25—informally, no fix-point subformula depends on an outer fix-point) then cl $\Phi \leq \omega$.

Proof. A straightforward structural induction shows that Φ is continuous in Z, that is, if $S_0 \subseteq S_1 \subseteq \cdots$ is an increasing chain of sets of states, then $\|\Phi\|_{\mathcal{V}[Z:=\cup_{i<\omega} S_i]}^\mathcal{T} = \bigcup_{i<\omega} \|\Phi\|_{\mathcal{V}[Z:=S_i]}^\mathcal{T}$, from which the result follows. □

For a simple example of a closure ordinal greater than ω, consider the following transition system:

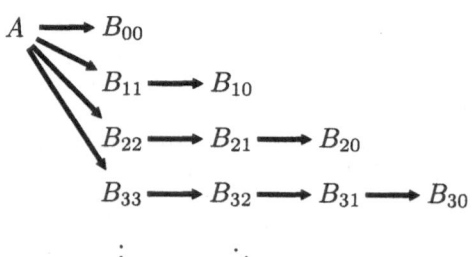

and the formula $\Phi \stackrel{\text{def}}{=} \mu Z.[-]Z$, which expresses 'eventually termination'. The denotation of Φ^k for $k < \omega$ is $\{\, B_{ij} \mid j < k \,\}$; A is not in any Φ^k since there is no bound on the length of paths from A. But Φ^ω comprises all the B_{ij} states, so A does satisfy $\Phi^{\omega+1}$.

Even if we do not use nested fix-point formulae, the mu-calculus is already powerful enough to express many of the logics mentioned earlier, so before going on to consider nested fix-points, we exhibit translations of PDL and CTL into the mu-calculus.

Definition 2.19. Given a PDL model $(\mathcal{W}, (\mathcal{V}_{\text{Atom}}, \mathcal{V}_{\text{Act}}))$, we define a mu-calculus model $(\mathcal{T}, \mathcal{V})$ of sort Act by making

$$\mathcal{T} = (\mathcal{W}, \{\, \mathcal{V}_{\text{Act}}(a) \mid a \in \text{Act}\,\})$$

and embedding Atom in Var and extending $\mathcal{V}_{\text{Atom}}$ to a valuation $\mathcal{V}$ on Var.

We can now define a translation $\text{Tr}_{\text{PDL-}\mu}$ (Tr in the next few paragraphs) of PDL formulae to mu-formulae thus:

$$\text{Tr}(P) = P$$

$$\text{Tr}(\neg\Phi) = \neg\text{Tr}(\Phi)$$

$$\text{Tr}(\Phi_1 \wedge \Phi_2) = \text{Tr}(\Phi_1) \wedge \text{Tr}(\Phi_2)$$

$$\text{Tr}(\langle a\rangle\Phi) = \langle a\rangle\text{Tr}(\Phi)$$

$$\text{Tr}(\langle\alpha_1;\alpha_2\rangle\Phi) = \text{Tr}(\langle\alpha_1\rangle\langle\alpha_2\rangle\Phi)$$

$$\text{Tr}(\langle\alpha_1 \cup \alpha_2\rangle\Phi) = \text{Tr}(\langle\alpha_1\rangle\Phi) \vee \text{Tr}(\langle\alpha_2\rangle\Phi)$$

$$\text{Tr}(\langle\alpha^*\rangle\Phi) = \mu Z.\text{Tr}(\Phi) \vee \text{Tr}(\langle\alpha\rangle Z)$$

If we have PDL with tests, we may add

$$\text{Tr}(\langle\Psi?\rangle\Phi) = \text{Tr}(\Psi) \wedge \text{Tr}(\Phi)$$

◁

Proposition 2.20. $\|\Phi\|^{\mathcal{W}}_{(\mathcal{V}_{\text{Atom}}, \mathcal{V}_{\text{Act}})} = \|\text{Tr}_{\text{PDL-}\mu}(\Phi)\|^{\mathcal{T}}_{\mathcal{V}}$

Proof. By induction on Φ. The only potential difficulty is in the proof of the inclusion $\supseteq$ for the case $\langle\alpha^*\rangle\Phi$, for while it is clear that $\|(\text{Tr}(\langle\alpha^*\rangle\Phi))^n\|$ is equal to $\|\Phi\| \cup \|\langle\alpha\rangle\Phi\| \cup \cdots \cup \|\langle\alpha^{n-1}\rangle\Phi\|$, it might be that $\text{Tr}(\langle\alpha^*\rangle\Phi)$ did not close at ω; however it does, since it is easy to see that $\text{Tr}(\langle\alpha\rangle Z)$ is continuous in Z. □

Certain special cases of PDL formulae occur often in applications (unsurprisingly), and are more easily understood in their PDL form than in their mu-calculus form. They are therefore prime candidates for 'macro formulae' in any practical use of the mu-calculus (a topic we discuss later). In particular, the PDL formulae

$$\langle a^* \rangle \Phi$$

$$[a^*] \Phi$$

meaning 'on some a-path' and 'on all a-paths' are common, and would be useful abbreviations for their mu-calculus translations

$$\mu Z. \Phi \vee \langle a \rangle Z$$

$$\nu Z. \Phi \wedge [a] Z$$

Turning now to CTL, the translation is also quite easy:

Definition 2.21. A CTL model $(\mathcal{T}, \mathcal{V})$ is viewed as a mu-calculus model by embedding Atom in Var and taking any singleton as the label set. We then define a translation $\mathrm{Tr}_{\mathrm{CTL}\text{-}\mu}$ from CTL formulae to mu-formulae as follows:

$$\mathrm{Tr}(P) = P$$

$$\mathrm{Tr}(\neg \Phi) = \neg \mathrm{Tr}(\Phi)$$

$$\mathrm{Tr}(\Phi_1 \wedge \Phi_2) = \mathrm{Tr}(\Phi_1) \wedge \mathrm{Tr}(\Phi_2)$$

$$\mathrm{Tr}(\exists \bigcirc \Phi) = \langle - \rangle \mathrm{Tr}(\Phi)$$

$$\mathrm{Tr}(\forall \bigcirc \Phi) = [-] \mathrm{Tr}(\Phi) \wedge \langle - \rangle \mathrm{tt}$$

$$\mathrm{Tr}(\exists (\Phi_1 \mathbf{U} \Phi_2)) = \mu Z. \Phi_2 \vee (\Phi_1 \wedge \langle - \rangle Z)$$

$$\mathrm{Tr}(\forall (\Phi_1 \mathbf{U} \Phi_2)) = \mu Z. \Phi_2 \vee (\Phi_1 \wedge \langle - \rangle Z \wedge \langle - \rangle \mathrm{tt})$$

◁

Note the presence of the clause $\langle - \rangle \mathrm{tt}$ in the translations of $\forall (\Phi_1 \mathbf{U} \Phi_2)$ and $\forall \bigcirc \Phi$. This is because the temporal operator $\Phi_1 \mathbf{U} \Phi_2$ is strong, in that it requires Φ_2 to be attained, and the temporal $\forall$ quantifies over all runs including the run of length zero if appropriate, whereas the modal $[-] \Phi$ does not require Φ to be attained—a state with no successors satisfies $[-]\mathrm{ff}$. As it happens, studies of CTL have usually required the successor relation to be total, in which case the $\langle - \rangle \mathrm{tt}$ clause is vacuous, but it is required if one allows termination.

Proposition 2.22. $\|\Phi\|_\nu^T = \|\mathrm{Tr}_{\mathrm{CTL}-\mu}(\Phi)\|_\nu^T$

Proof. Again a mostly routine induction; the only slight difficulty is that although CTL formulae are defined on states, the temporal operators in terms of which they are expressed, are defined in terms of runs, whereas the mu-calculus makes no mention of paths. At the end of the next chapter, we shall see how the tableau system shows this equivalence very simply.

$\square$

The formulae we have exhibited so far have barely scratched the surface of the mu-calculus; by embedding fix-point formulae in other fix-point formulae, we can consider more complex properties. Before doing so, we establish what we mean by 'embedding', for the notion is more subtle than the straightforward subformula relation. The difference we must capture is that between

$$\nu Y.(\mu Z.P \vee [-]Z) \wedge [-]Y$$

and

$$\nu Y.\mu Z.(P \vee [-]Z) \wedge [-]Y.$$

The first formula contains a minimal fix-point within a maximal fix-point, but the inner fix-point does not depend on the outer, whereas in the second formula it does. There is a lack of good terminology for this distinction— the second concept is called 'alternation' by Emerson and Lei [EmL86] who first defined it, but this term could equally apply to the first concept. However, we shall follow Emerson and Lei, but also define specific terms for each concept, together with a notation for alternation depth based on a natural analogy with predicate logic: we shall use (light-face) σ-alternation for the first notion, and (bold-face) $\boldsymbol{\sigma}$-alternation for the second, with the understanding that plain 'alternation' means '$\boldsymbol{\sigma}$-alternation'.

Definition 2.23. The classes $\hat{\mu}_{\leq 0}$ and $\hat{\nu}_{\leq 0}$ of mu-formulae are equal, and are the class of formulae containing no σ-formulae as subformulae. For $i \geq 0$, the class $\hat{\mu}_{\leq i+1}$ (resp. $\hat{\nu}_{\leq i+1}$) is the class of formulae Φ such that every ν-subformula $\nu Z.\Psi$ (resp. μ-subformula $\mu Z.\Psi$) is in the class $\hat{\nu}_{\leq i}$ (resp. $\hat{\mu}_{\leq i}$). The class $\hat{\delta}_{\leq i}$ is $\hat{\mu}_{\leq i} \cap \hat{\nu}_{\leq i}$.

A formula Φ is in the class $\hat{\mu}_n$ iff n is the least i such that Φ is in $\hat{\mu}_{\leq i}$.

The σ-*alternation depth* (where σ is a symbol *per se*) of a formula Φ is the greatest n such that some fix-point subformula of Φ is in $\hat{\sigma}_n$ (where σ is μ or ν). $\triangleleft$

To anticipate, unnested formulae of the classes $\hat{\mu}_i$, $\hat{\nu}_i$ and $\hat{\delta}_i$ are analogous to the classes Σ_i, Π_i and Δ_i of first order predicate logic. The classes $\hat{\mu}_{\leq n}$

etc. are a messy technicality caused by the absence of a prenex normal form—note that $\hat{\mu}_{\leq n} = \hat{\mu}_n \cup \hat{\mu}_{\leq n-1} \cup \hat{\nu}_{\leq n-1}$.

The notion of σ-alternation is slightly tricky to define. The original definition in [EmL86] was by a set of inductive rules; our definition, although equivalent, is cast in terms chosen to relate it as closely as possible to σ-alternation. It is necessary to introduce the notion of 'subsentence', which is a little more complex than usual since we have forsaken the formal distinction between variables and atomic propositions, but there are compensations elsewhere.

Definition 2.24. For a formula Φ, a fix-point subformula $\sigma Z.\Psi$ is a *fix-subsentence* of Φ if for every variable X in Ψ, either X is bound in $\sigma Z.\Psi$ or X is free in Φ.

The *intrinsic class* of a σ-formula $\sigma Z.\Phi$ is defined thus: let Φ' be the result of replacing every fix-subsentence of Φ by a variable symbol P that does not occur in $\sigma Z.\Phi$; then if $\sigma Z.\Phi'$ is in the class $\hat{\sigma}_n$, the intrinsic class of $\sigma Z.\Phi$ is σ_n. A σ-formula has intrinsic class $\sigma_{\leq n}$ if it has intrinsic class σ_n or μ_i or ν_i for some $i < n$.

The class $\mu_{\leq n}$ (resp. $\nu_{\leq n}$) of mu-formulae is the class of formulae Φ such that every fix-subsentence of Φ has intrinsic class $\mu_{\leq n}$ (resp. $\nu_{\leq n}$); $\delta_{\leq n} = \mu_{\leq n} \cap \nu_{\leq n}$; and μ_n etc. are defined as for the corresponding lightface classes. Then the (σ-)alternation depth of a formula is the greatest n such that some fix-point subformula has intrinsic class σ_n (equivalently, the least n such that the formula is not in class $\delta_{\leq n}$).   ◁

The motivation for the above terminology is that the class of a formula measures the highest alternation depth encountered anywhere in that formula, whereas the intrinsic class of a σ-formula measures the complexity of that particular fix-point, ignoring accidental complexity due to other fix-point sub-sentences. We can now define the term 'unnested' which we have already used:

Definition 2.25. A σ-formula is an *unnested fix-point* iff all its proper fix-point subformulae are fix-subsentences. It is an *alternating fix-point* iff it is in the class σ_n for some $n > 1$.

A formula is in μ_n (etc.) if it is in $\hat{\mu}_n$ and is unnested.   ◁

Note that being unnested is a more stringent condition than being unalternating: $\mu Y.\mu Z.[-]Y \vee \langle - \rangle Z$ is neither unnested nor alternating.

Some examples of the application of these definitions are:

- $\Phi \stackrel{\text{def}}{=} \mu Y.(\nu Z.P \wedge [-]Z) \vee \langle - \rangle Y$

The intrinsic class of $\nu Z.P \wedge [-]Z$ is $\boldsymbol{\nu_1}$, and that of Φ is $\boldsymbol{\mu_1}$ (since the ν-subsentence is eliminated by substitution), so Φ has alternation depth 1 (and, for what it is worth, Φ is in $\boldsymbol{\delta_2}$).

- $\Phi \stackrel{\text{def}}{=} \nu Y.\mu Z.[-]Y \vee \langle - \rangle Z$
  The intrinsic class of $\mu Z.[-]Y \vee \langle - \rangle Z$ is $\boldsymbol{\mu_1}$, and of Φ is $\boldsymbol{\nu_2}$ (since the μ-subformula is not a μ-subsentence), and so Φ has class $\boldsymbol{\nu_2}$.

- $\Phi \stackrel{\text{def}}{=} \mu X.\nu Y.\mu Z.[-]Z \vee \langle - \rangle Y \vee \langle a \rangle X$
  Similarly, this is in $\boldsymbol{\mu_3}$.

- $\Phi \stackrel{\text{def}}{=} \mu X.\nu Y.(\mu Z.[-]Z \vee \langle - \rangle Y) \vee \langle a \rangle X$
  This is a potential trick: at first sight, one might think that this had alternation depth 2, since the inner fix-point only mentions Z and Y but not X; however, since Y depends on X, the formula is in fact in $\boldsymbol{\mu_3}$, as can be seen from applying the rule.

What sort of properties can be expressed with alternating fix-points? As mentioned earlier, the alternation of fix-points is as subtle as the alternation of quantifiers—one of the interesting empirical facts about the mu-calculus is that until recently no computationally natural properties with an alternation depth greater than three had been found.

At alternation depth two, the major properties are those concerned with infinite frequency of behaviour; perhaps the simplest alternating property is 'infinitely often P'

$$\Phi \stackrel{\text{def}}{=} \nu Y.\mu Z.\Phi' \stackrel{\text{def}}{=} \nu Y.\mu Z.(P \vee [-]Z) \wedge [-]Y \wedge \langle - \rangle \text{tt}.$$

Let us consider why this formula means what we claim. The easiest approach to understanding fix-point formulae is one due to Stirling, in which the idea is to consider repeatedly unfolding the fix-point formula as one traverses some path through the system—but a least fix-point may only be unfolded finitely often, whereas one may unfold greatest fix-points without limit. This idea is the key to the next chapter—the conditions for success of tableaux formalize it, and I have found that the intuitive idea and the tableau formalization reinforce one another effectively, increasing one's comprehension of the mu-calculus. If we apply this idea to Φ, we see that

$$s \vDash \Phi$$

iff $s \vDash \mu Z.\Phi'[\Phi/Y]$

iff $s \vDash (P \vee [-]\mu Z.\Phi'[\Phi/Y]) \wedge [-]\Phi \wedge \langle - \rangle \text{tt};$

now, we see that for one thing, all successors of s also satisfy Φ (and so the argument we are conducting applies to all states reachable from s), and further, either $s \vDash P$ or its successors satisfy $\mu Z.\Phi'[\Phi/Y]$; in the latter case we unfold the least fix-point, and find that the same argument applies to these successors—the important point is that we may only unfold the least fix-point finitely often, so that within finite time we must hit a state satisfying P—and then we start over again. Thus the formula says that P holds within finite time, and when it does hold, Φ still holds in the next states, and so P holds infinitely often.

We have introduced Φ as an example of an alternating fix-point; yet it may seem that the property 'infinitely often P' is expressible by a non-alternating fix-point, namely

$$\Psi \stackrel{\text{def}}{=} \nu Y.(\mu Z.P \vee ([-]Z \wedge \langle-\rangle\text{tt})) \wedge [-]Y \wedge \langle-\rangle\text{tt}$$

which says 'always (eventually $(P \wedge \langle-\rangle\text{tt}))$'.† It so happens that the properties 'infinitely often' and 'always eventually' are equivalent when immediately governed by a universal branching quantifier, and so Φ and Ψ are equivalent. However, this is not the case if the quantification is existential; but before considering this, let us continue exploring the meaning of Ψ. An alternative to thinking of unfoldings is to think of the approximants. This is usually more difficult, but for simpler formulae it can give interesting insights. So, consider the approximants of Ψ and Φ.

It should be clear that Ψ^i means '(eventually P) holds at all states reachable in less than i steps', and so the limit is 'always eventually P'.

Now consider the approximants of Φ:

$$\Phi = \nu Y.\mu Z.(P \vee [-]Z) \wedge [-]Y \wedge \langle-\rangle\text{tt}$$

$$\Phi^0 = \text{'all states'}$$

---

† The reader may be a little confused by the appearance and non-appearance of $\langle-\rangle\text{tt}$'s in various places. In the mu-calculus examples, we use as few of these as possible: Φ is intended to be a translation of the temporal property $\forall \mathbf{GF}(P \wedge \langle-\rangle\text{tt})$; the explicit $\langle-\rangle\text{tt}$ there is required because the temporal $\mathbf{G}$ operator permits termination, whereas we do not intend to permit termination when we say 'infinitely often'; then another $\langle-\rangle\text{tt}$ is added by the translation of $\mathbf{F}$ (see Definition 2.21, recalling that $\mathbf{F}P \stackrel{\text{def}}{=} \text{tt } \mathbf{U} P$), thus giving a 'natural' translation of $\nu Y.\mu Z.((P \wedge \langle-\rangle\text{tt}) \vee ([-]Z \wedge \langle-\rangle\text{tt})) \wedge [-]Y$, which simplifies to Φ. (The word 'natural' is in quotes because the translation of PTL formulae more complex than CTL into the mu-calculus is a very tricky subject, recently addressed by [Dam90] for CTL*.)

$$\Phi^1 = \mu Z.(P \vee [-]Z) \wedge [-]\Phi^0 \wedge \langle - \rangle \text{tt}$$

$$= \text{'eventually } P\text{'}$$

$$\Phi^2 = \mu Z.(P \vee [-]Z) \wedge [-]\Phi^1 \wedge \langle - \rangle \text{tt}$$

$$= \text{'eventually } P, \text{ and upto and including } P, \text{ next(eventually } P)\text{'}$$

Thus Φ^i says 'P happens at least i times on all paths from here', so the limit is truly 'infinitely often P'.

We can demonstrate the difference between 'infinitely often' and 'always eventually' with the existential formulae

$$\Phi \stackrel{\text{def}}{=} \nu Y.\mu Z.(P \vee \langle - \rangle Z) \wedge \langle - \rangle Y$$

$$\Psi \stackrel{\text{def}}{=} \nu Y.(\mu Z.P \vee \langle - \rangle Z) \wedge \langle - \rangle Y$$

which express the temporal properties $\exists \mathbf{GF}(P \wedge \langle - \rangle \text{tt})$ and $\exists \mathbf{G} \exists \mathbf{F}(P \wedge \langle - \rangle \text{tt})$, the first of which says 'there exists a path along which P holds infinitely often', and the second of which says 'there exists a path along which P is always attainable'; for example, in the system

where $\|P\| = \{B\}$, Φ fails everywhere, but because of the path $A \longrightarrow A \longrightarrow A \longrightarrow \cdots$, Ψ is true at A.

So far, all our examples have used, in effect, the unlabelled mu-calculus, so we should explain the benefits of the labelled calculus, and particularly of our use of sets of labels. The reasons for using labels are well-known: it allows one to express properties such as

$$\mu Z.P \vee ([a]Z \wedge \langle a \rangle \text{tt})$$

meaning 'P holds eventually on all a-paths' (where a path σ is an a-path iff $\forall i \leq |\sigma| . \mathcal{L}(\sigma, i) = a$) and other properties distinguishing paths by their events. What then is the reason for extending to sets of labels? Clearly if the sort is finite, as is usually the case, using sets of labels adds nothing to the expressive power of the language (since $[K] = \bigwedge_{a \in K}[a]$), but it does allow the *concise* expression of interesting properties, particularly by means of the $[-K]$ notation. These properties express the idea of *events happening*, rather than *states being*. The simplest example is

$$\Phi \stackrel{\text{def}}{=} \mu Z.\langle-\rangle\text{tt} \wedge [-a]Z$$

which means '*a* happens eventually'. We may see this either by unfolding: we may only pass through the Z finitely often, so there are no infinite $-a$-runs, and the $\langle-\rangle$tt ensures that we can't terminate before doing an a; or by approximating: Φ^1 is '*a* happens now' and Φ^i is '*a* happens within $i-1$ steps'.

Compare Φ with

$$\Psi \stackrel{\text{def}}{=} \mu Z.([-a]\text{ff} \vee [-]Z) \wedge \langle-\rangle\text{tt}$$

$$\Upsilon \stackrel{\text{def}}{=} \mu Z.\langle a\rangle\text{tt} \vee ([-]Z \wedge \langle-\rangle\text{tt})$$

Ψ means 'eventually a must happen', i.e. we always reach a point where a is the only event possible, and Υ means 'eventually a may happen', i.e. we always reach a point where a is possible. The distinction between these three formulae is illustrated by the system

in which Φ holds at A, B and D, Ψ holds only at D, and Υ holds at A, B, C, D and F.

By combining label sets with alternation, complex fairness properties may be expressed. For example, one might wish to say 'on all paths, either a happens or b happens infinitely often', which is true of A and B in the system

This is expressed by the formula

$$\nu Y.\mu Z.[-a,b]Z \wedge [b]Y \wedge \langle-\rangle\text{tt}.$$

Intuitively, paths are only allowed finitely many non-a events, but a b event resets the counter to zero.

With higher alternation depths one can express complex cyclic properties while allowing for CCS-style divergence. For example, the following formula of alternation depth four says that the visible behaviour of a CCS process must match the regular expression $(ab + abc)^*$, where the process may not diverge after the a:

$$\mu W.\nu X.[-a, \tau]\text{ff} \wedge [\tau]X \wedge [a]\mu Y.[-b, \tau]\text{ff} \wedge [\tau]Y$$

$$\wedge\; [b]\nu Z.[-a, c, \tau]\text{ff} \wedge [\tau]Z \wedge [c]W \wedge [a]Y$$

Chapter 3

The Tableau System

In subsection 1.2.5 we described briefly the model checking technique used by Emerson, Clarke *et al.* An important feature of this technique is that it is 'global', that is, it constructs the entire state graph of the system and then traverses it—and in the original version [CES86] the system was traversed once for every subformula of the formula being checked. Clearly this causes difficulties as soon as systems become large, let alone infinite; but since one is more usually interested in whether a particular state satisfies a formula than whether such a state exists, this global traversal may not even be necessary, since the truth of a particular property at a particular state may depend only on a small neighbourhood of that state. Thus there are good reasons to look at 'local' model-checking, where checking that a property holds at a state is done by considering only the local behaviour of the system, so far as necessary.

Stirling and Walker [StW89] provided a method of local model-checking for the mu-calculus which used a 'tableau method'. Tableau methods have long been used for establishing validity etc. (see, for example, [Fit83]), and in [Sti87] Stirling advocated the use of tableau methods for showing relative truth, as in model-checking.

This monograph is about the extension of tableau methods to infinite systems. There are several reasons why this should be done. Firstly, while in practice all implementations are of course finite, few people would wish to give up using potentially infinite models. Secondly, if we have a method for dealing with infinite systems, *a fortiori* we have a method for dealing with large systems. The third reason is more general: a belief that computer-*assisted* reasoning, as opposed to automated reasoning, has an important part to play in program verification. The tableau system presented here is certainly not intended to be completely automated, but it does, I believe, provide a sensible demarcation between that which should be automated and that which must be left to a human. Why is this so? The mu-calculus allows, as we have mentioned more than once, the expression of immensely complex properties, which in general are far from decidable, and may require powerful theories for their proof. But a major cause of difficulty is least fix-points, expressing termination and the like; the tableau method is in some sense a device for separating the easy part of model-checking, namely boolean and modal connectives, from the difficult part, namely least fix-point checking. A computer should deal with the easy part, but the human prover is left free to apply whatever techniques are needed to prove the difficult part.

This method brings advantages even to small finite model-checking—the original Stirling–Walker model-checker is not suitable for naive implementation as it stands, owing to a very high worst-case complexity (k-tuply exponential for formulae of alternation depth k, as opposed to $|\mathcal{T}|^{k+1}$ for Emerson and Lei's [EmL86] global algorithm); now one could of course develop ways of reusing information efficiently, as Cleaveland did for his variant tableau method [CPS89] when he implemented it in the Edinburgh Concurrency Workbench, but the generalized tableau does this anyway, if sensibly used, simply by taking sets of states rather than single states as the objects to be checked. Of course, it remains possible to produce enormous tableaux for simple problems, but this should not happen with an intelligent user.

Before describing the tableau system formally, I shall attempt to motivate it intuitively, for although the details appear complex, the basic idea is simple.

3.1 Intuition behind the tableau system.

The system is goal-directed, that is, we have a statement of the form $S \models \Phi$ (where S is a set of states) that we wish to prove true, so we start with a sequent $S \vdash \Phi$ and apply natural deduction style rules to obtain subgoals according to the structure of Φ. The rules for boolean and modal connectives are fairly straightforward—to prove $S \vdash \Phi_1 \wedge \Phi_2$ we must prove $S \vdash \Phi_1$ and $S \vdash \Phi_2$, and to prove $S \vdash \langle K \rangle \Phi$ we must for each $s \in S$ find a K-successor s', and then prove $S' \vdash \Phi$ for the resulting set of successors. To deal with fix-points we take what is effectively an unfolding approach, where infinite unfolding is avoided by incorporating implicitly a way of looping back to an earlier point in the tableau. So, for example, if we were to start proving $S \vdash \nu Z.P \wedge [-]Z$, we might unfold a few times and get to a point where we were trying to prove $S' \vdash \nu Z.P \wedge [-]Z$ for some $S' \subseteq S$—but then we could 'loop back' to the beginning, where we were trying to prove the same formula for the bigger set S. In the case of greatest fix-points, this could also be thought of as fix-point induction—if we can prove the fix-point for S by assuming it for a smaller set S', that is a proof for S. However, for least fix-points it is necessary to exercise careful control of such looping, for two reasons. Firstly, recall that for a least fix-point we may only unfold finitely often. This is fundamental to the proof of least fix-points, and is dealt with by imposing a termination condition outside the tableau proper. The second reason is that we cannot just unfold blindly, because of the following problem: consider

$$\mu Y.\nu Z.[-]Z \wedge [-]Y.$$

Suppose we unfold the outer fix-point: we get

$$\nu Z.[-]Z \wedge [-]\mu Y.\nu Z.[-]Z \wedge [-]Y.$$

We then naturally proceed to unfold the outer ν, and then the μ, and then the second (which has now become the outer) ν—but the second ν is distinct from the first, and we must not allow 'looping back' from the second to the first: To see why this is so, we can consider approximants: if we wish to show that $s \vDash \mu Y. \dots$, we can try to show that there is some $\alpha + 1$ such that $s \vDash \mu^{\alpha+1} Y. \dots$. Now if we unfold this, we get

$$\nu Z.[-]Z \wedge [-]\mu^{\alpha} Y. \dots$$

so that the first ν we unfold is relative to μ^{α}, and the second to $\mu^{\alpha'}$ for some $\alpha' < \alpha$ and so on; so we cannot use the first to prove the second.

This problem is dealt with by introducing so-called propositional constants, following Stirling and Walker's original system. The purpose of such constants is just to allow us to distinguish between on the one hand repeated unfoldings of the same formula, and on the other, distinct appearances of the formula, as above. Instead of unfolding fix-point formulae directly, when we encounter a fix-point for the first time, we allocate a fresh constant to that formula, replacing the fix-point variable by the constant, and then unfold the constant. So in the example above, we first allocate a constant U to the outer μ, and when we unfold we get

$$\nu Z.[-]Z \wedge [-]U$$

and then allocating a new constant V and unfolding again

$$[-]V \wedge [-]U.$$

If we now elect to unfold the U, we get

$$[-]V \wedge [-](\nu Z.[-]Z \wedge [-]U)$$

and so the new occurrence of ν is assigned a different constant V', and is not confused with the first. (One could choose to think of constant introduction as a kind of incremental preservation of normal form.)

Using the rules, we construct a tableau. We then have to show that the tableau is 'successful', that is, gives a correct proof of its root sequent. Intuitively, this means (a) checking atomic propositions and (b) checking that 'loops' obey the termination conditions adumbrated above. It is in (b) that (perhaps very strong) reasoning outside the tableau is needed, since we do this by proving well-foundedness of a relation on states which is defined by the tableau, a relation that encapsulates the idea of the truth of a fix-point at s depending on the truth of the fix-point at another state s'.

3.2 Definition of the tableau system.

Definition 3.1. Let U, V, W range over a (countably infinite) set of *propositional constants*, and extend the mu-calculus to permit constants in formulae.

A *definition* is $U = \Phi$ where Φ is an (extended) formula.

A *definition list* is a finite (possibly empty) sequence $\Delta = (U_1 = \Phi_1, \ldots, U_n = \Phi_n)$ of definitions such that the U_i are distinct and the only constants appearing in Φ_i are those in $\{U_1, \ldots, U_{i-1}\}$. We use $\Delta \cdot (U = \Phi)$ to denote the definition list $(U_1 = \Phi_1, \ldots, U_n = \Phi_n, U = \Phi)$, provided that this satisfies the necessary conditions. If $U = \Phi$ is a definition in Δ then $\Delta(U)$ is Φ.

Given a model $(\mathcal{T}, \mathcal{V})$, the interpretation of a formula Φ with respect to a definition list Δ is $\|\Phi_\Delta\|_\mathcal{V}^\mathcal{T}$ defined by the following inductive rule:

$$\|\Phi_{\Delta \cdot (U=\Psi)}\|_\mathcal{V} = \|\Phi_\Delta\|_{\mathcal{V}[U := \|\Psi_\Delta\|_\mathcal{V}]}.$$

◁

We now define the sequents of our system: as well as a set of states and a formula of the mu-calculus with constants, sequents carry with them a definition list giving the meanings of any constants in the formula. Henceforth we assume some fixed model $(\mathcal{T}, \mathcal{V})$.

Definition 3.2. A sequent has the form $S \vdash_\Delta \Phi$ where S is a set of states, Φ is a formula of the extended mu-calculus in positive normal form, and Δ is a definition list.   ◁

Definition 3.3. A tableau rule is one of the following, where the premise appears above the consequents.

$\wedge$
$$\frac{S \vdash_\Delta \Phi_1 \wedge \Phi_2}{S \vdash_\Delta \Phi_1 \qquad S \vdash_\Delta \Phi_2}$$

$\vee$
$$\frac{S \vdash_\Delta \Phi_1 \vee \Phi_2}{S_1 \vdash_\Delta \Phi_1 \qquad S_2 \vdash_\Delta \Phi_2}$$

where $S = S_1 \cup S_2$

$[K]$
$$\frac{S \vdash_\Delta [K]\Phi}{S' \vdash_\Delta \Phi}$$

where $S' = \{\, s' \mid \exists s \in S, a \in K \,.\, s \xrightarrow{a} s' \,\}$

$\langle K \rangle$
$$\frac{S \vdash_\Delta \langle K \rangle \Phi}{f(S) \vdash_\Delta \Phi}$$

where $f \colon S \to f(S)$ is a function such that $\forall s \in S \,.\, \exists a \in K \,.\, s \xrightarrow{a} f(s)$

$\sigma Z.$
$$\frac{S \vdash_\Delta \sigma Z.\Phi}{S \vdash_{\Delta'} U}$$

where U is a constant not appearing in Δ, and $\Delta' = \Delta \cdot (U = \sigma Z.\Phi)$

Un
$$\frac{S \vdash_\Delta U}{S \vdash_\Delta \Phi[U/Z]}$$

where $\Delta(U) = \sigma Z.\Phi$

Thin
$$\frac{S \vdash_\Delta \Phi}{S' \vdash_\Delta \Phi}$$

where $S' \supseteq S$.   ◁

The mention of the function f in the $\langle K \rangle$ rule should be noted. The purpose of this function is to choose the successor state which most quickly leads towards 'termination' (cf. the choice function used in [StE89])—we require that this function be specified explicitly when the rule is applied. Choosing this function intelligently is essential: choosing the wrong successor may result in a tableau failing even though it is trying to prove a true property. This is inevitable, since the point of this system is to reduce the generally intractable problem of checking a formula to something which can be reasoned about, and some knowledge of the system under investigation is required to do this. The other rules whose intelligent application is necessary are Thin and $\vee$; judicious application of Thin makes tableaux finite, and $\vee$ requires a choice just as $\langle K \rangle$ does. It is perhaps somewhat inconsistent to permit S_1 and S_2 to intersect in the $\vee$ rule; however, allowing all relevant successors in a $\langle K \rangle$ would complicate the definition of success so much as to vitiate the system entirely, whereas allowing intersecting disjuncts just makes it more likely that unsuccessful tableaux are produced. It is, therefore, recommended that in practice S_1 and S_2 are always made disjoint.

Definition 3.4. A *tableau* is a proof-tree built from a *root* sequent $S_0 \vdash_{()} \Phi_0$, where Φ_0 is a mu-formula, by application of the above rules, repeated until all leaves of the tree are terminal. A leaf node $\mathbf{n} = S \vdash_\Delta \Phi$ is *terminal* if no rule other than Thin applies, that is

(i) $\Phi = Z$ or $\Phi = \neg Z$, or

(ii) $\Phi = \langle K \rangle \Psi$ and $\exists s \in S . \forall a \in K, s' \in S . \text{not } s \xrightarrow{a} s'$,

or the state set is trivial

(iii) $S = \varnothing$

or finally

(iv) $\Phi = U$ and $\Delta(U) = \sigma Z.\Psi$ and $\mathbf{n}$ has an ancestor node $\mathbf{n}' = S' \vdash_{\Delta'} U$ such that $S' \supseteq S$.

A node fulfilling (iv) is called a *σ-terminal*. ◁

Before going on to consider when tableaux are successful, we record a few obvious facts.

Proposition 3.5. For a sequent $S \vdash_\Delta \Phi$ labelling a node $\mathbf{n}$ in a tableau

(i) Δ is indeed a definition list;

(ii) every definition in Δ has the form $U = \sigma Z.\Psi$;

(iii) for any sequent $S' \vdash_{\Delta'} \Phi'$ labelling an ancestor node $\mathbf{n}'$ of $\mathbf{n}$, Δ is an extension of Δ';

(iv) any constant appearing in Φ is defined in Δ;

(v) the interpretation of Φ with respect to Δ, $\|\Phi_\Delta\|$, accords with the natural syntactic definition

$$\|\Phi_{()}\|_\mathcal{V}^\mathcal{T} = \|\Phi\|_\mathcal{V}^\mathcal{T}$$

$$\|\Phi_{\Delta \cdot (U = \Psi)}\|_\mathcal{V}^\mathcal{T} = \|\Phi[\Psi/U]_\Delta\|_\mathcal{V}^\mathcal{T}.$$

Proof. (i)-(iv) are completely trivial. (v) is also a routine induction; the only point worth noticing is that since in a tableau free variables never appear unless they are free in the root sequent, the substitution does not even have to worry about α-converting to avoid variable capture. $\square$

For a tableau to provide a valid proof of its root, the leaves of the tableau must be true.

Definition 3.6. A terminal node as in Definition 3.4 is *successful* if
- it is in class (i) and is true, i.e. if $\Phi = Z$ then $S \subseteq \mathcal{V}(Z)$ and if $\Phi = \neg Z$ then $S \cap \mathcal{V}(Z) = \varnothing$; or
- it is in class (iii); or
- it is in class (iv) with $\sigma = \nu$; or
- it is in class (iv) with $\sigma = \mu$ and satisfies the mu-success conditions defined below;

otherwise it is unsuccessful.

A tableau is successful if it is finite and all its terminals are successful.

$\triangleleft$

The condition for success of least fix-point terminals is unavoidably complex, and proceeds via several auxiliary definitions.

Definition 3.7. Let $\mathbf{n}$ be a σ-terminal. The ancestor node $\mathbf{n}'$ in Definition 3.4 (iv) is called the *companion* of $\mathbf{n}$; if there is more than one such node, the companion is the lowest.

$\triangleleft$

As will be seen from the soundness proof, if there is more than one such ancestor, it does not actually matter which is taken to be the companion; however, taking the lowest makes the success condition easier to evaluate in practice.

We now define a notion of path through the tableau, which formalizes the intuitive ideas we gave earlier for understanding fix-point formulae.

Definition 3.8. In a tableau, a *path* from a state s at a node $\mathbf{n}$ to a state s' at a node $\mathbf{n}'$ is a sequence $(s, \mathbf{n}) = (s_0, \mathbf{n}_0), (s_1, \mathbf{n}_1), \ldots, (s_k, \mathbf{n}_k) = (s', \mathbf{n}')$ of states and nodes such that
- $\mathbf{n}_{i+1}$ is a child of $\mathbf{n}_i$;

- if $\mathbf{n}_i = S_i \vdash_{\Delta_i} \Phi_i$ then $s_i \in S_i$;
- if the rule applied to $\mathbf{n}_i$ is $[K]$ then $s_i \xrightarrow{K} s_{i+1}$, if the rule is $\langle K \rangle$ then $s_{i+1} = f(s_i)$, and otherwise $s_{i+1} = s_i$.

We write $s@\mathbf{n} \longrightarrow s'@\mathbf{n}'$ if there is a path from s at $\mathbf{n}$ to s' at $\mathbf{n}'$.

Now there is an *extended path* from s at $\mathbf{n}$ to s' at $\mathbf{n}'$, written $s@\mathbf{n} \dashrightarrow s'@\mathbf{n}'$, if either

(i) $s@\mathbf{n} \longrightarrow s'@\mathbf{n}'$, or

(ii) there is a node $\mathbf{n}'' = S'' \vdash_{\Delta''} U$ and a finite sequence of states $s_0, s_1, \ldots, s_k$ and nodes $\mathbf{n}_1, \ldots, \mathbf{n}_k$ for $k \geq 0$, where each $\mathbf{n}_i$ is a terminal with companion $\mathbf{n}''$, such that $s@\mathbf{n} \longrightarrow s_0@\mathbf{n}''$ and $s_i@\mathbf{n}'' \dashrightarrow s_{i+1}@\mathbf{n}_{i+1}$ for $0 \leq i < k$, and $s_k@\mathbf{n}'' \dashrightarrow s'@\mathbf{n}'$. ◁

Note that the definition of extended path is recursive; it is well-defined because an extended path from $\mathbf{n}$ is defined in terms of extended paths from strict descendants of $\mathbf{n}$.

These extended paths are the same as the 'trails' of [BrS90]—the definitions are different in each case to suit the proof techniques.

Definition 3.9. Let $\mathbf{n} = S \vdash_\Delta U$ be a μ-terminal with companion node $\mathbf{n}' = S' \vdash_{\Delta'} U$. Define a relation $\sqsupseteq_{\mathbf{n}'}$ on S' by $s \sqsupseteq_{\mathbf{n}'} s'$ iff $s@\mathbf{n}' \dashrightarrow s'@\mathbf{n}''$ for any terminal $\mathbf{n}''$ whose companion is $\mathbf{n}'$. The condition mu-success is that $\sqsubseteq_{\mathbf{n}'}$ should be well-founded, that is, there should be no infinite chains $s_0 \sqsupseteq_{\mathbf{n}'} s_1 \sqsupseteq_{\mathbf{n}'} \cdots$. ◁

Thus the success of μ-terminals is really a feature of their companions, not of the individual terminals.

The main result is now that the tableau system is sound and complete:

Theorem 3.10. In a model $(\mathcal{T}, \mathcal{V})$, there is a successful tableau with root $S \vdash_{()} \Phi$ if and only if $S \subseteq \|\Phi\|_{\mathcal{V}}^{\mathcal{T}}$.

Proof. Deferred until after some examples: see Theorem 3.13 for soundness, and Theorem 3.23 for completeness. □

3.3 Simple examples.

We now consider some very simple examples of tableaux to demonstrate the rules. We take the examples of mu-formulae in chapter 2, and exhibit tableaux for these formulae on the example systems of that chapter.

The simplest fix-point formula we looked at was $\mu Z.[-]Z$ on the transition system

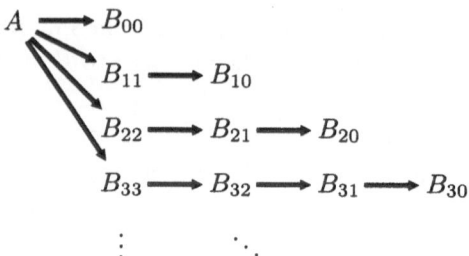

We should like to prove that $A \in \|\mu Z.[-]Z\|$. A tableau for this is

$$
\begin{array}{l}
_1\,\{A\} \vdash_{()} \mu Z.[-]Z \\
\hline
\qquad _2\,\{A\} \vdash_{\Delta} U \\
\hline
_3\,\{A, B_{ij}\} \vdash_{\Delta} U \\
\hline
_4\,\{A, B_{ij}\} \vdash_{\Delta} [-]U \\
\hline
\qquad _5\,\{B_{ij}\} \vdash_{\Delta} U
\end{array}
\qquad \Delta = (U = \mu Z.[-]Z)
$$

This already demonstrates the essential use of Thin at node 2: if we did not include all the B states there, we would never stop unfolding. To show that the terminal is successful, we must show that $\sqsubset_3$ is well-founded; now $\sqsupset_3$ is exactly the $\longrightarrow$ relation, and there are no infinite $\longrightarrow$ paths—which is what we are trying to prove!

For a less trivial example, consider the existential 'always eventually' formula

$$
\Psi \stackrel{\text{def}}{=} \nu Y.(\mu Z.P \vee \langle - \rangle Z) \wedge \langle - \rangle Y
$$

on the system

$$
\circlearrowleft \; A \longrightarrow B \longrightarrow C \; \circlearrowright
$$

where P holds at B as before. The proof that A satisfies Ψ is

$$\frac{\{A\} \vdash_0 \nu Y.(\mu Z.P \vee \langle - \rangle Z) \wedge \langle - \rangle Y}{\{A\} \vdash_\Delta U} \quad \Delta = (U = \nu Y....)$$

$$\frac{}{\{A\} \vdash_\Delta (\mu Z.P \vee \langle - \rangle Z) \wedge \langle - \rangle U}$$

$$\frac{\{A\} \vdash_\Delta \mu Z.P \vee \langle - \rangle Z \wedge \langle - \rangle U}{2\,\{A\} \vdash_{\Delta'} V} \quad \Delta' = \Delta \cdot (V = \mu Z....) \qquad \frac{1\,\{A\} \vdash_\Delta \langle - \rangle U}{\{A\} \vdash_\Delta U}$$

$$\frac{}{3\,\{A,B\} \vdash_{\Delta'} V}$$

$$\frac{\{A,B\} \vdash_{\Delta'} P \vee \langle - \rangle V}{\{B\} \vdash_{\Delta'} P \qquad 4\,\{A\} \vdash_{\Delta'} \langle - \rangle V}$$

$$\frac{}{5\,\{B\} \vdash_{\Delta'} V}$$

This tableau illustrates several points. We must specify the functions f_1 and f_4 used in the diamond rules at nodes 1 and 4—obviously we chose $f_1(A) = A$ and $f_4(A) = B$. We also used Thin at node 2 in order to make 3 be a companion to 5 (and so the relation $\sqsupset_3$ is $\{(A,B)\}$ which is well-founded). By making these choices we have produced the shortest tableau (a 'canonical' tableau as defined in section 3.5). However, we need not have been successful so quickly. The choice of f_1 is essential—the other possibility would produce an unsuccessful tableau—but the use of Thin is not; in fact, Thin is never necessary for finite systems. If we did not use Thin, but just kept on unfolding, we would get a tableau that agrees with the first as far as node 2, and then looks like

$$\frac{\{A\} \vdash_{\Delta'} V}{\{A\} \vdash_{\Delta'} P \vee \langle - \rangle V}$$

$$\frac{\{\} \vdash_{\Delta'} P \qquad \{A\} \vdash_{\Delta'} \langle - \rangle V}{\{B\} \vdash_{\Delta'} V}$$

$$\frac{\{B\} \vdash_{\Delta'} P \vee \langle - \rangle V}{\{B\} \vdash_{\Delta'} P \qquad \{\} \vdash_{\Delta'} \langle - \rangle V}$$

which does not even have any μ-terminals to consider.

As for the diamond function in this part of the tableau, if we make the wrong choice, the tableau will fail; but one could relax the termination conditions (by allowing, rather than forcing, termination at a terminal) to allow the wrong choice to be made at first, as long as the right choice is made eventually.

Finally, we should consider an example of an alternating fix-point. Recall that in the above transition system no state satisfies the existential 'infinitely often' formula

$$\Phi \overset{\text{def}}{=} \nu Y.\mu Z.(P \vee \langle - \rangle Z) \wedge \langle - \rangle Y.$$

Let us prove this by showing that all states satisfy the negation of Φ, namely

$$\mu Y.\nu Z.(\neg P \wedge [-]Z) \vee [-]Y.$$

A tableau (again canonical) for this is

$$\cfrac{\cfrac{\cfrac{\cfrac{\{A,B,C\} \vdash_{()} \mu Y.\nu Z.(\neg P \wedge [-]Z) \vee [-]Y}{{}_1\{A,B,C\} \vdash_\Delta U} \quad \Delta = (U = \mu Y....)}{\{A,B,C\} \vdash_\Delta \nu Z.(\neg P \wedge [-]Z) \vee [-]U}}{{}_2\{A,B,C\} \vdash_{\Delta'} V} \quad \Delta' = \Delta \cdot (V = \nu Z....)}{\{A,B,C\} \vdash_{\Delta'} (\neg P \wedge [-]V) \vee [-]U}}{}$$

$$\cfrac{\{A,C\} \vdash_{\Delta'} (\neg P \wedge [-]V)}{\{A,C\} \vdash_{\Delta'} \neg P \qquad \cfrac{\{A,C\} \vdash_{\Delta'} [-]V}{{}_3\{B,C\} \vdash_{\Delta'} V}} \qquad \cfrac{\{B\} \vdash_{\Delta'} [-]U}{{}_4\{C\} \vdash_{\Delta'} U}$$

For this to be successful, we must have that $\sqsubset_1$ is well-founded. Since this is an alternating fix-point, $\sqsupset_1$ depends also on $\sqsupset_2$. If we follow paths down directly from 1 to the terminal 4, we get $B \sqsupset_1 C$; but at 2 we have $A \sqsupset_2 B$ and $C \sqsupset_2 C$, so when we incorporate this we get also that $A \sqsupset_1 C$; this still leaves $\sqsubset_1$ well-founded, so our tableau is successful.

3.4 Soundness of the tableau system.

The proof of soundness proceeds by a somewhat subtle induction on the height of sub-tableaux. This proof is based on the 'approximant' view of fix-points; contrast Stirling's proof in [BrS90], which uses the unfolding approach more directly (but still, of course, uses approximants).

We have the problem that sub-trees of tableaux are not themselves tableaux; therefore our first step is to extend the notion of tableau so we can do induction on the height of tableaux. It is also necessary to allow definition lists to bind constants to approximants; this corresponds to the use of ν-signatures in [BrS90].

Definition 3.11. Let Δ be a definition list with $\Delta(W) = \sigma^\gamma Z.\Phi$ for some $\gamma \in \text{Ord} \cup \{\infty\}$. $\Delta[W:\gamma']$ denotes the definition list which agrees with Δ except that $\Delta[W:\gamma'](W) = \sigma^{\gamma'} Z.\Phi$.

A *quasi-tableau* is a proof tree built from a root sequent $S_0 \vdash_{\Delta_0} \Phi_0$, where Δ_0 may bind ν-constants to fix-points or approximants; the rules applied are those for tableaux; and the termination conditions are those for tableaux with the following addition

(v) if W is defined in the root definition list Δ_0, a node $S \vdash_\Delta W$ that has no companion may be a leaf, or Un may be applied; if it is a leaf, it is called a *quasi-terminal*.

A quasi-tableau is successful if all its terminals are successful. (The notion of successful quasi-tableau is a generalization of the valuation-independent successful tableaux of [Bra91].)

If τ is a quasi-tableau and $\mathbf{n}$ is a node of τ, $\tau(\mathbf{n})$ denotes the quasi-tableau formed by the sub-tree of τ rooted at $\mathbf{n}$.

If τ is a quasi-tableau with root definition list Δ such that $\Delta(V) = \nu^\beta Y.\Psi$, $\tau[V:\beta']$ denotes the quasi-tableau formed from τ by replacing every definition list Δ' by $\Delta'[V:\beta']$.   ◁

Proposition 3.12. Let τ be a successful quasi-tableau with root $\mathbf{n} = S \vdash_\Delta \Phi$. Suppose the root is false, that is, there exists a state s such that $s \notin \|\Phi_\Delta\|$. Then there exists a quasi-terminal $\mathbf{n}' = S' \vdash_{\Delta'} W'$ and a state $s' \in S'$ such that $s' \notin \|W'_{\Delta'}\|$ and

(i) if the rule applied to $\mathbf{n}$ is not Un, then $s@\mathbf{n} \twoheadrightarrow s'@\mathbf{n}'$

(ii) if the rule applied to $\mathbf{n}$ is Un, then there is a finite sequence of states $s = s_0, s_1, \ldots, s_k$ for some $k \geq 0$ such that $s_i \sqsupset_{\mathbf{n}} s_{i+1}$ for $i < k$ and $s_k@\mathbf{n} \twoheadrightarrow s'@\mathbf{n}'$.

Proof. The proof is by induction on the height of the quasi-tableau. The base case, for quasi-tableaux of height 2, is (almost) trivial, and is left to the reader. (We cannot take height 1 as the base case, since the definition of extended path does not allow for a terminal being its own companion.)

We consider the possibilities for the rule applied to $\mathbf{n}$.

Suppose the rule is $\wedge$, so that $\Phi = \Phi_1 \wedge \Phi_2$. By definition of the semantics, $s \notin \|\Phi_\Delta\|$ if $s \notin \|\Phi_{1\Delta}\|$ or $s \notin \|\Phi_{2\Delta}\|$ and so there is a path from s to a child (possibly both) $\mathbf{n}''$ of $\mathbf{n}$. Now apply the proposition inductively to s and $\tau(\mathbf{n}'')$. If case (i) holds, we are done, since $s@\mathbf{n} \twoheadrightarrow s@\mathbf{n}''$ and $s@\mathbf{n}'' \twoheadrightarrow s'@\mathbf{n}'$ implies $s@\mathbf{n} \twoheadrightarrow s'@\mathbf{n}'$; and if case (ii) holds, again we are done since (by the definition of $\sqsupset_{\mathbf{n}''}$) we have exactly part (ii) of the definition of $s@\mathbf{n} \twoheadrightarrow s'@\mathbf{n}'$ (Definition 3.8).

A similar argument applies to the rules $\vee$, $[K]$, $\langle K \rangle$, Thin and $\sigma Z.$, so we have now only to deal with the unfolding rule.

Suppose that $\Phi = U$ with $\Delta(U) = \mu X.\Psi$, so the child node is $\mathbf{n}'' = S \vdash_\Delta \Psi[U/X]$. Now $s \in \|U_\Delta\|$ iff $s \in \|\Psi[U/X]_\Delta\|$ by definition of fix-points, so apply the proposition inductively to s and $\tau(\mathbf{n}'')$. Just as above, we have $s@\mathbf{n} \twoheadrightarrow s'@\mathbf{n}'$, and $\mathbf{n}'$ is a quasi-terminal in $\tau(\mathbf{n}'')$. Now either $\mathbf{n}'$ is also a quasi-terminal in τ, in which case we are done, or $\mathbf{n}$ is the companion of $\mathbf{n}'$. In the latter case, we have $s \sqsupset_\mathbf{n} s'$, so set $s_1 = s'$ and repeat the argument for s_1 and τ. Either the repetition process terminates, giving us the desired sequence of states, or we have an infinite sequence $s_0 \sqsupset_\mathbf{n} s_1 \sqsupset_\mathbf{n} \cdots$, which contradicts the well-foundedness of $\sqsubset_\mathbf{n}$.

Finally, suppose that $\Phi = V$ with $\Delta(V) = \nu^\beta Y.\Psi$ for some $\beta \in \mathrm{Ord} \cup \{\infty\}$, so the child is $\mathbf{n}'' = S \vdash_\Delta \Psi[V/Y]$. By definition, if $s \notin \|\nu^\beta Y.\Psi_\Delta\|$ there is a least ordinal, which must be a successor ordinal, say $\beta_0 + 1$, such that $s \notin \|\nu^{\beta_0+1} Y.\Psi_\Delta\|$ (and so $\beta_0 < \beta$), and therefore that $s \notin \|\Psi[V/Y]_{\Delta[V:\beta_0]}\|$. Apply the proposition inductively to s and $\tau(\mathbf{n}'')[V:\beta_0]$; we get a quasi-terminal $\mathbf{n}' = S' \vdash_{\Delta'[V:\beta_0]} W$ of $\tau(\mathbf{n}'')[V:\beta_0]$ and $s' \in S$ such that $s' \notin \|W_{\Delta'[V:\beta_0]}\|$; but by monotonicity we have that $s' \notin \|W_{\Delta'[V:\beta]}\|$. So if $\mathbf{n}'$ is a quasi-terminal in τ we are done. Otherwise, $\mathbf{n}$ is the companion of $\mathbf{n}'$, so we set $s_1 = s'$ and repeat the argument with s_1 and $\tau[V:\beta_0]$. In this case, if the process does not terminate, giving us the desired sequence, we get an infinite sequence $\beta_0 > \beta_1 > \cdots$, which is impossible. $\qquad\square$

Theorem 3.13. If $S \vdash_{()} \Phi$ is the root of a successful tableau τ, then $S \subseteq \|\Phi\|$.

Proof. A tableau is a special case of a quasi-tableau, so by the proposition if the root is false there is a quasi-terminal; but a successful tableau has no quasi-terminals. $\qquad\square$

3.5 Completeness of the tableau system.

The completeness proof is somewhat easier, since it is not necessary to worry about multiple unfoldings—we build a tableau in which the constants are unfolded only once.

We define a notion of μ-signature which records how low down the approximant hierarchy a state satisfies a least fix-point. The components of the signature play a role dual to that of the ordinals β used in the soundness proof. (These signatures are not quite the same as the signatures used in

Streett and Emerson's decidability proof: their signatures have components for μ-subsentences, whereas we need to deal with all μ-subformulae.)

Definition 3.14. Let $\mathbf{n} = S \vdash_\Delta \Phi$ be a node in a tableau, and let $U_1, U_2, \ldots, U_k$ enumerate the μ-constants in Δ in order of appearance. The *signature* $\text{sig}(s, \mathbf{n})$ at $\mathbf{n}$ of a state s is the lexicographically least sequence of ordinals $(\alpha_1, \ldots, \alpha_k)$ such that $s \in \|\Phi_{\Delta[U_1:\alpha_1]\ldots[U_k:\alpha_k]}\|$, if such a sequence exists. Signatures exist for all states in a true node, by Proposition 2.13. ◁

Lemma 3.15. If in the above definition $\Phi = U_i$, then α_i is a successor ordinal and $\alpha_j = 0$ for $j > i$.
Proof. Immediate. □

Proposition 3.16. Let S and Φ_0 be such that $S \subseteq \|\Phi_0\|$. Build a proof tree from the sequent $S_0 \vdash_{()} \Phi_0$ by repeated application of the following procedure:

Choose a non-terminal leaf node $\mathbf{n} = S \vdash_\Delta \Phi$.

- If the top level connective of Φ is boolean or modal, apply the relevant rule to create children $\mathbf{n}_1$ and $\mathbf{n}_2$. If the rule is $\vee$ choose the sets labelling the children thus: $S_1 = \{\, s \in S \mid \text{sig}(s, \mathbf{n}_1) \leq \text{sig}(s, \mathbf{n}_2) \,\}$ and $S_2 = S - S_1$. If the rule is $\langle K \rangle$, so $\Phi = \langle K \rangle \Psi$, define the function f by letting $f(s)$ be any s' such that $\forall s'' \in \|\Psi_\Delta\| . s \xrightarrow{K} s'' \Rightarrow \text{sig}(s'', \mathbf{n}_1) \geq \text{sig}(s', \mathbf{n}_1)$.
- If Φ is a fix-point formula, apply σZ.
- If $\Phi = W$, first apply Thin, labelling the resulting node with the set $S' = \|W_\Delta\|$, and then apply Un.

This procedure terminates and produces a tableau in which every node is true, and moreover Un is applied exactly once to each constant.

Proof. Directly from the semantics, if the procedure is called on a true node it creates true nodes. If $\mathbf{n} = S \vdash_\Delta W$ is a leaf node, and Un has already been applied to some node $\mathbf{n}' = S' \vdash_{\Delta'} W$ higher up the tree, then $S' = \|W_{\Delta'}\|$, and since both nodes are true, $S \subseteq S'$, so $\mathbf{n}$ is terminal with companion $\mathbf{n}'$. □

Definition 3.17. Such a tableau is called *canonical*. ◁

Lemma 3.18. If $s@\mathbf{n} \longrightarrow s'@\mathbf{n}'$ is a path in the above tableau from the premise to a conclusion of any rule other than μX. or Un applied to a μ-constant, then $\text{sig}(s', \mathbf{n}') \leq \text{sig}(s, \mathbf{n})$.
Proof. For the $[K]$ rule, from the semantics the signature of s at $\mathbf{n}$ is the supremum of the signatures of its successors at $\mathbf{n}'$, and similarly for the $\wedge$ rule.

For the $\langle K \rangle$ and $\vee$ rules, the construction of the tableau explicitly chose the successor with least signature, and so from the semantics $\text{sig}(s, \mathbf{n}) = \text{sig}(s', \mathbf{n}')$.

Thin and $\nu Y.$ are trivial.

For Un applied to a ν-constant V with $\Delta(V) = \nu Y.\Psi$, we need only note that $s \in \|V_{\Delta[U_1:\alpha_1]...[U_k:\alpha_k]}\|$ iff $s \in \|\Psi[V/Y]_{\Delta[U_1:\alpha_1]...[U_k:\alpha_k]}\|$, so the signature remains unchanged. $\qquad\square$

Lemma 3.19. If $s@\mathbf{n} \longrightarrow s@\mathbf{n}'$ is a path across a $\mu X.$ rule, and $\text{sig}(s, \mathbf{n}) = (\alpha_1, \ldots, \alpha_k)$, then $\text{sig}(s, \mathbf{n}') = (\alpha_1, \ldots, \alpha_k, \alpha_{k+1})$ for some α_{k+1}.
Proof. Immediate. $\qquad\square$

Lemma 3.20. If $s@\mathbf{n} \longrightarrow s@\mathbf{n}'$ is a path across Un applied to a μ-constant, then $\text{sig}(s, \mathbf{n}') < \text{sig}(s, \mathbf{n})$.
Proof. Constants are unfolded only immediately after their introduction, so $\text{sig}(s, \mathbf{n})$ has the form $(\alpha_1, \ldots, \alpha_{k-1}, \alpha + 1)$ where $\mathbf{n} = S \vdash_\Delta U_k$. By definition of approximant, $\text{sig}(s, \mathbf{n}') = (\alpha_1, \ldots, \alpha_{k-1}, \alpha)$. $\qquad\square$

Lemma 3.21. If $\mathbf{n} = S \vdash_\Delta U$ is a companion node, and $s \sqsupset_\mathbf{n} s'$, then $\text{sig}(s, \mathbf{n}) > \text{sig}(s', \mathbf{n})$.
Proof. If $\text{sig}(s, \mathbf{n}) = (\alpha_1, \ldots, \alpha_k)$, by the previous lemma, the signature decreases across the Un rule applied to $\mathbf{n}$, and by the other lemmas the first k components of the signature do not increase along the extended path from s at $\mathbf{n}$ to s' at some U-terminal. $\qquad\square$

Corollary 3.22. $\sqsubset_\mathbf{n}$ is well-founded. $\qquad\square$

Theorem 3.23. If $S \subseteq \|\Phi\|$, the tableau constructed by the above procedure applied to the root sequent $S \vdash_{()} \Phi$ is successful.
Proof. All terminals are true, and all μ-terminals satisfy mu-success, from the Corollary above. $\qquad\square$

3.6 Variations on the theme.

We consider now some variations of the system, most of which are aimed at reducing the size of tableaux.

If we wish to ensure that tableau construction terminates, we can do so by forcing termination at a certain depth. We define the tableau system of *degree* k by adding the constraint that a node $\mathbf{n} = S \vdash_\Delta U$ such that Un has been applied k times to U above $\mathbf{n}$ is terminal. Since canonical tableaux are of degree 1, the system of degree k is sound and complete, for

$k \geq 1$. In the case $k = 1$ we can dispense with the propositional constants entirely: such a system was presented in [Bra91].

A different change allows us to recapture the original [StW89] system for finite systems: restrict the sequent sets to be singletons, omit the Thin rule, and replace the $[K]$ rule by

$$\frac{\{s\} \vdash_\Delta [K]\Phi}{\{s_1\} \vdash_\Delta \Phi \quad \cdots \quad \{s_n\} \vdash_\Delta \Phi}$$

where $\{s_1, \ldots, s_n\} = \{ s' \mid s \xrightarrow{K} s' \}$. In this case all μ-terminals **n** are unsuccessful as there must be an extended path from the single state labelling the companion **n**′ of **n** to the same state at **n**, giving a cyclic $\sqsubset_{\mathbf{n}'}$.

The Thin rule is a good candidate for elimination: as shown by the completeness proof, it need only be applied to constants, so it could be incorporated into Un. On the other hand, we could instead allow implicit thinning at every rule application, which would eliminate the common annoyance where one has a sequent $S \vdash_\Delta [K]V$, and all one needs is that the successor set of S should be contained in a companion set to V, but one has to calculate the exact successor set to write it down in the tableau.

There are many derived operators and rules that may be convenient: we give a few examples.

Finite n-ary conjunction and disjunction with the obvious n-consequent rules are particularly useful for the schematic tableau we introduce in the next chapter.

An Aristotelian 'for all' operator is a useful abbreviation, particularly when considering fairness properties—the reader will recall the frequent occurrence of $\langle - \rangle$tt. Since this operator will be used later, we define it officially:

Definition 3.24. $[K]$ is a derived mu-calculus operator given by $[K]\Phi \overset{\text{def}}{=} [K]\Phi \wedge \langle K \rangle$tt. It has a derived tableau rule

$$[K] \qquad \frac{S \vdash_\Delta [K]\Phi}{S' \vdash_\Delta \Phi}$$

where $S' = \{ s' \mid \exists s \in S . s \xrightarrow{K} s' \}$ and furthermore $\forall s \in S . \exists s' . s \xrightarrow{K} s'$. The operator $\langle K \rangle$ is defined dually. ◁

We may wish to use CTL formulae as abbreviations, since they express common properties. To give derived rules for these, we can extend the rules dealing with constants to deal with each CTL formula as well as the mu-calculus fix-points; for example, for $\exists\mathbf{G}\Phi$ we would have the constant introduction rule

$$\frac{S \vdash_\Delta \exists\mathbf{G}\Phi}{S \vdash_{\Delta'} U}$$

where $\Delta' = \Delta \cdot (U = \exists\mathbf{G}\Phi)$, and an additional variant of the unfolding rule:

$$\frac{S \vdash_\Delta U}{S \vdash_\Delta \langle\!-\!\rangle U \qquad S \vdash_\Delta \Phi}$$

where $\Delta(U) = \exists\mathbf{G}\Phi$, together with the condition that U is treated as a ν-constant.

Now that we have mentioned CTL formulae as abbreviations for mu-formula, it is a convenient point to demonstrate how the tableau system allows a very easy proof of the correctness of the translation which justifies such abbreviation. Consider, for example, $\exists(\Phi_1 \ \mathbf{U} \ \Phi_2)$. The translation is $\mu Z.\Phi_2 \vee (\Phi_1 \wedge \langle-\rangle Z)$, and a tableau for the denotation S of this has the form

$$\frac{\dfrac{\dfrac{\dfrac{S \vdash_{()} \mu Z.\Phi_2 \vee (\Phi_1 \wedge \langle-\rangle Z)}{S \vdash_\Delta U}}{S \vdash_\Delta \Phi_2 \vee (\Phi_1 \wedge \langle-\rangle U)}}{S_2 \vdash_\Delta \Phi_2 \qquad \dfrac{S_1 \vdash_\Delta \Phi_1 \wedge \langle-\rangle U}{S_1 \vdash_\Delta \Phi_1 \qquad \dfrac{S_1 \vdash_\Delta \langle-\rangle U}{f(S_1) \vdash_\Delta U}}}}{}$$

Now a sequence $s_0 \sqsupset s_1 \sqsupset \ldots \sqsupset s_n$ with $s_n \vDash \Phi_2$, which must exist by the well-foundedness condition, is exactly a path of the form required by the definition of $s_0 \vDash \exists(\Phi_1 \ \mathbf{U} \ \Phi_2)$, and conversely such a path can be used to build a successful tableau by taking S to be all the states in the path.

3.7 The tableau system and Hoare logic.

As promised earlier, we now show how Floyd–Hoare logic relates to the tableau system. The constructions are unfortunately rather ugly, but they do show how a Hoare proof can be transformed into a tableau system proof.

Take a standard simple **while**-language as in subsection 1.2.1. A Hoare proof system for partial correctness is as follows:

Skip $\qquad\qquad\qquad\qquad\{P\}\mathbf{skip}\{P\}$

Assignment $\qquad\qquad\qquad\{Q[e/x]\}x := e\{Q\}$

Sequence $\qquad\qquad\dfrac{\{P\}c_1\{R\}\quad\{R\}c_2\{Q\}}{\{P\}c_1;\ c_2\{Q\}}$

Conditional $\qquad\dfrac{\{P\wedge b\}c_1\{Q\}\quad\{P\wedge\neg b\}c_2\{Q\}}{\{P\}\mathbf{if}\ b\ \mathbf{then}\ c_1\ \mathbf{else}\ c_2\{Q\}}$

While $\qquad\qquad\dfrac{\{I\wedge b\}c\{I\}}{\{I\}\mathbf{while}\ b\ \mathbf{do}\ c\{I\wedge\neg b\}}$

Consequence $\qquad\dfrac{P\Rightarrow P'\quad\{P'\}c\{Q'\}\quad Q'\Rightarrow Q}{\{P\}c\{Q\}}$

We give a translation of partial correctness formulae into mu-formulae (actually, into sequents), and show how the above rules can be translated into rules for the construction of a successful tableau.

The translation works on weakest preconditions rather than partial correctness formulae; the pre-condition P in $\{P\}c\{Q\}$ appears on the left of the sequent.

Definition 3.25. Let $c\{Q\}$ denote the weakest precondition of a command c with respect to an assertion Q; that is, a state s satisfies $c\{Q\}$ iff any terminating execution of c at s results in a state satisfying Q. A translation Tr() from weakest preconditions to mu-formulae is defined by induction on c as follows:

$$\mathrm{Tr}(a\{Q\}) = [a]Q$$

$$\mathrm{Tr}(c_1;\ c_2\{Q\}) = \mathrm{Tr}(c_1\{\mathrm{Tr}(c_2\{Q\})\})$$

$$\mathrm{Tr}(\mathbf{if}\ b\ \mathbf{then}\ c_1\ \mathbf{else}\ c_2\{Q\}) = (b\wedge\mathrm{Tr}(c_1\{Q\}))\vee(\neg b\wedge\mathrm{Tr}(c_2\{Q\}))$$

$$\mathrm{Tr}(\mathbf{while}\ b\ \mathbf{do}\ c\{Q\}) = \nu Z.(\neg b\wedge Q)\vee(b\wedge\mathrm{Tr}(c\{Z\}))$$

◁

Proposition 3.26. Let $\{P\}$ (on the left of a sequent) denote the set of states satisfying the assertion P. Then $\{P\}c\{Q\}$ iff $\{P\} \vDash \text{Tr}(c\{Q\})$.

Proof. By induction on the structure of c. The only non-trivial case is the translation for **while**: if

$$\{P\}\textbf{while } b \textbf{ do } c\{Q\}$$

then the equivalent form

$$\{P\}\textbf{if } \neg b \textbf{ then skip else } (c; \textbf{ while } b \textbf{ do } c)\{Q\}$$

gives us

$$\{P\} \vDash (\neg b \wedge Q) \vee (b \wedge \text{Tr}(c\{\text{Tr}(\textbf{while } b \textbf{ do } c\{Q\})\}))$$

and so by fix-point induction

$$\{P\} \vDash \nu Z.(\neg b \wedge Q) \vee (b \wedge \text{Tr}(c\{Z\}))$$

and conversely; if $\{P\}$ satisfies the fix-point, unfolding once gives us the **if** expansion in Hoare logic.   □

So to prove $\{P\}c\{Q\}$ it suffices to build a successful tableau for $\{P\} \vdash_{()} \text{Tr}(c\{Q\})$. Given a Hoare proof of $\{P\}c\{Q\}$, we can construct a tableau by following the Hoare proof.

Proposition 3.27. Suppose we have a proof tree in Hoare logic for the assertion $\{P\}c\{Q\}$. Construct a tableau by the following inductive procedure according to the Hoare rule applied to $\{P\}c\{Q\}$ (in each case, assume $\{P\}c\{Q\}$ is as given in the above list of rules, and the premises are as in the list of rules). During the inductive construction, the tableau may have an initial definition list Δ; this is empty at the top level.

Skip. The tableau is

$$\{P\} \vdash_\Delta P$$

Assignment. The tableau is

$$\frac{\{Q[e/x]\} \vdash_\Delta [x := e]Q}{\{Q\} \vdash_\Delta Q}$$

Sequence. Let τ_1 be the tableau constructed for $\{P\}c_1\{R\}$. The tableau is formed by replacing R by $\text{Tr}(c_2\{Q\})$ in τ_1; then for each leaf of the form $\{R\} \vdash_{\Delta'} \text{Tr}(c_2\{Q\})$, replace it by the tableau τ_2 for $\{R\}c_2\{Q\}$.

Conditional. The tableau is

$$\frac{\{P\} \vdash_\Delta (b \wedge \mathrm{Tr}(c_1\{Q\})) \vee (\neg b \wedge \mathrm{Tr}(c_2\{Q\}))}{\begin{array}{cc} \dfrac{\{P \wedge b\} \vdash_\Delta b \wedge \mathrm{Tr}(c_1\{Q\})}{\{P \wedge b\} \vdash_\Delta b \qquad \tau_1} & \dfrac{\{P \wedge \neg b\} \vdash_\Delta \neg b \wedge \mathrm{Tr}(c_2\{Q\})}{\{P \wedge \neg b\} \vdash_\Delta \neg b \qquad \tau_2} \end{array}}$$

where τ_1 is the tableau for $\{P \wedge b\}c_1\{Q\}$ and τ_2 that for $\{P \wedge \neg b\}c_2\{Q\}$.
While. The tableau is

$$\frac{\dfrac{\{I\} \vdash_\Delta \nu Z.(\neg b \wedge I) \vee (b \wedge \mathrm{Tr}(c\{Z\}))}{\{I\} \vdash_{\Delta'} V}}{\dfrac{\{I\} \vdash_{\Delta'} (\neg b \wedge I) \vee (b \wedge \mathrm{Tr}(c\{V\}))}{\begin{array}{cc} \dfrac{\{I \wedge \neg b\} \vdash_{\Delta'} \neg b \wedge I}{\{I \wedge \neg b\} \vdash_{\Delta'} \neg b \quad \{I \wedge \neg b\} \vdash_{\Delta'} I} & \dfrac{\{I \wedge b\} \vdash_{\Delta'} b \wedge \mathrm{Tr}(c\{V\})}{\{I \wedge b\} \vdash_{\Delta'} b \qquad \tau} \end{array}}}$$

where τ is the tableau for $\{I \wedge b\}c\{I\}$ with every appearance of I on the right of sequents replaced by V.
Consequence. The tableau is

$$\frac{\{P\} \vdash_\Delta \mathrm{Tr}(c\{Q'\})}{\tau}$$

where τ is the tableau for $\{P'\}c\{Q'\}$ with Q' replaced by Q on the right-hand side and then each leaf $\{Q'\} \vdash_{\Delta'} Q$ thinned to $\{Q\} \vdash_{\Delta'} Q$.

Then the resulting tableau is successful.

Proof. By inspection, every rule applied is applied correctly; for example, the rule in the translation of assignment is valid because in the Hoare model assignment can always be done, so any state satisfying Q can be reached by doing $x := e$ from some state satisfying $Q[e/x]$. The key observation is that this construction ensures that after any stage of the construction any terminal whose right hand side is the Q of $\{P\}c\{Q\}$ is actually $\{Q\} \vdash_{\Delta'} Q$ (some Δ'), and so all the V-terminals in the While clause are of the form $\{I\} \vdash_{\Delta'} V$ and so are successful. $\qquad\square$

Thus the soundness of Hoare logic follows from the soundness of the tableau system.

Relative completeness of the Hoare system is of course dependent on the expressibility of weakest preconditions in the assertion language. Given this, the completeness of the tableau system implies the completeness of

Hoare logic: all we do is construct the canonical tableau for $\{P\} \vdash_{()}$ $\mathrm{Tr}(c\{Q\})$ and run the above translation backwards. (Without the expressibility, we could still produce a canonical tableau, but the backwards translation might not work since the sets of states in sequents might not be describable by formulae of the Hoare assertion language.)

Finally, this goes over directly to total correctness. The only difference when dealing with total correctness is that the ν in the translation of **while** formulae is replaced by μ. The Floyd approach for proving termination is, as has been mentioned, to find a well-founded relation on the state space that decreases as execution passes through the body of a **while**-loop; but an execution path through the body of a loop corresponds exactly to an extended path in the tableau from the introduction of the corresponding V constant to a V-terminal (the looping through subordinate fix-point nodes reflecting nested **while**-loops), so such a Floyd ordering serves as a $\sqsupset$ relation in the tableau, and conversely a tableau $\sqsupset$ relation is a Floyd ordering.

Chapter 4

Applications to Nets

This chapter is devoted to applications of the tableau system, demonstrating some of the techniques that can be used to build tableaux for systems, and showing how the arithmetical nature of Petri nets allows attractive arithmetical reasoning about tableaux.

We start with an introductory section defining Petri nets, illustrating ways of building systems from nets, and mentioning some interesting properties of nets, and then proceed to applications of the tableau system to nets.

4.1 Petri nets.

4.1.1 Basic definitions.

Petri nets are commonly represented in graphical form, thus:

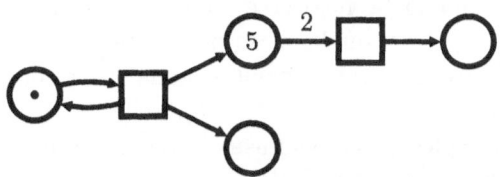

The circles represent *places* which may hold an arbitrary number of tokens (represented by black blobs, or by a number (e.g. 5, meaning there are 5 tokens)). The boxes represent *transitions* which fire, and when a transition fires, it removes a token from each place connected to it by an arc, and adds one to each place to which it is connected. Sometimes arcs have a weight (e.g. 2 in the example) giving the number of tokens transferred along the arc when a transition fires; omitted weights are always unity. (This informal description of net behaviour is known as the *token game*.) Nets are formalized thus:

Definition 4.1. A *place-transition net* $\mathcal{N} = (\mathbf{S}, \mathbf{T}; \mathbf{F})$ comprises two disjoint sets $\mathbf{S}$ (the *places*) and $\mathbf{T}$ (the *transitions*) together with a map $\mathbf{F} : (\mathbf{S} \times \mathbf{T}) \cup (\mathbf{T} \times \mathbf{S}) \to \mathsf{N}$ (where $\cup$ is disjoint union), the *flow relation* (strictly a multi-relation).

A *marking* of $\mathcal{N}$ is a map $M : \mathbf{S} \to \mathsf{N}$. Maps from a set X into N are identified with $|X|$-dimensional N-vectors in the natural way, and so $M(s)$ is often written M_s.

For $x \in \mathbf{S} \cup \mathbf{T}$, ${}^\bullet x$ is the map $\mathbf{S} \cup \mathbf{T} \to \mathsf{N}$ given by ${}^\bullet x(y) = \mathbf{F}(y, x)$, and similarly $x^\bullet(y) = \mathbf{F}(x, y)$; sometimes ${}^\bullet x$ will be abused to mean the set $\{\, y \mid {}^\bullet x(y) > 0 \,\}$, the pre-places (pre-transitions) of x, and similarly for post-places.

Nets may be equipped with initial markings, usually written M_0; in this thesis we do not consider an initial marking to be part of the definition of net.   ◁

Hereafter, *net* means place-transition net unless otherwise stated.

Notation 4.2. The use of bold is non-standard. In [Rei85] the multi-relation we write as $\mathbf{F}$ is split into a relation F and a weight function W assigning a weight to each $(x, y) \in F$; it seems simpler to work with multi-relations and allow the occasional abuse of notation.

We write $\beta : X \to_\mu Y$ to mean $\beta : Y \times X \to \mathsf{N}$, and identify multirelations with matrices in the natural way.

We shall sometimes use the term *event* for transition, particularly when nets are being mentioned in the same breath as transition systems. Strictly, the term 'event' should be restricted to the discussion of *condition/event systems*, which may be thought of as nets with capacities (see Definition 4.5) with all arc weights and capacities equal to 1, but the confusion is mostly harmless.

In specific examples, places will usually be named by roman majuscules, and transitions by roman minuscules.

A net called $\mathcal{N}$ has the form $(\mathbf{S}, \mathbf{T}; \mathbf{F})$ unless otherwise stated.   ◁

The token game is formalized by the following definitions.

Definition 4.3. For a net $\mathcal{N}$, a transition t is *M-enabled*, or enabled at M, iff ${}^\bullet t \leq M$.

M' is a *successor marking* of M via t, written $M \xrightarrow{t} M'$, iff t is M-enabled and $M' = M - {}^\bullet t + t^\bullet$.

M' is *reachable* from M if there exist $M = M_1, t_1, M_2, t_2, \ldots, t_{n-1}$, $M_n = M'$ such that M_{i+1} is a successor via t_i of M_i for $1 \leq i < n$.   ◁

Notation 4.4. We write $M \xrightarrow{t} M'$; the standard notation is $M[t\rangle M'$, which we avoid in order to reduce confusion with the modal logic operators, and because we mostly view nets as transition systems, where the $\xrightarrow{t}$ notation is standard. We use the following derived notations: $M \longrightarrow M'$ means that $\exists t \in \mathbf{T}.\, M \xrightarrow{t} M'$; $M \xrightarrow{t}$ means $\exists M'.\, M \xrightarrow{t} M'$; and the symbol $\overrightarrow{M}$ denotes $\{\, M' \mid M \longrightarrow^* M' \,\}$, the set of markings reachable from M, where * is the Kleene star.   ◁

The above assumes implicitly that only one transition may fire at a time. Much of the intuitive appeal of nets arises from their non-interleaving semantics, where many transitions can fire concurrently. This is formalized as above, replacing the single transition t by a non-empty set T (or even a multiset) of transitions and defining $^{\bullet}T = \sum_{t \in T} {}^{\bullet}t$. However, as stated earlier, we mostly use an interleaving semantics.

Definition 4.5. A *net with capacities* is a net $\mathcal{N}$ equipped with a function $K: \mathbf{S} \to \mathbb{N} \cup \{\infty\}$. In such a net the definition of M-enabled is modified to $(^{\bullet}t \leq M) \wedge (M + t^{\bullet} \leq K)$.   ◁

Thus the capacity $K(s)$ of a place is the maximum number of tokens it can hold. We shall use nets with capacities *ad libitum* in presenting examples, but shall regard them merely as an abbreviation for a plain net, via the standard complementing construction: for a net with capacities $(\mathcal{N}, K)$ we build a net $\mathcal{N}'$ thus: let $\mathbf{S}_{\text{fin}} = \{\, s \in \mathbf{S} \mid K(s) < \infty \,\}$, and let $\overline{\mathbf{S}}_{\text{fin}}$ be a disjoint copy of $\mathbf{S}_{\text{fin}}$. Then $\mathcal{N}' = (\mathbf{S}', \mathbf{T}; \mathbf{F}')$ where $\mathbf{S}' = \mathbf{S} \cup \overline{\mathbf{S}}_{\text{fin}}$, and $\mathbf{F}'(s, t)$ is $\mathbf{F}(s, t)$ for $s \in \mathbf{S}$ and $\mathbf{F}(t, s)$ for $s \in \overline{\mathbf{S}}_{\text{fin}}$, and similarly for $\mathbf{F}'(t, s)$. A marking M of $(\mathcal{N}, K)$ gives a marking M' of $\mathcal{N}'$ by $M'(s) = M(s)$ for $s \in \mathbf{S}$ and $M'(s) = K(s) - M(s)$ for $s \in \overline{\mathbf{S}}_{\text{fin}}$. It is easily seen that the behaviour of $\mathcal{N}'$ from M' exactly simulates that of $(\mathcal{N}, K)$ from M, since for each place s in $\mathcal{N}$, in $\mathcal{N}'$ the sum $M(s) + M(\overline{s})$ remains constant at $K(s)$.

4.1.2 Properties and classes of nets.

One of the most basic and important notions in net analysis is that of invariant.

Definition 4.6. A $(S\text{-})$*invariant* of a net $\mathcal{N}$ is a $|\mathbf{S}|$-dimensional $\mathbb{Z}$-vector ι such that for any marking M and successor marking M' of M, $\iota \cdot M = \iota \cdot M'$, where $\cdot$ is the usual scalar product of vectors.   ◁

So an invariant is a linear combination of places whose value is constant throughout the behaviour of the net. This is conveniently expressed in linear algebraic terms: define a $|\mathbf{S}| \times |\mathbf{T}|$ matrix N by $N_{st} = \mathbf{F}(s, t) - \mathbf{F}(t, s)$;

then, viewing markings and (multi)sets of transitions as column vectors, for $M \xrightarrow{t} M'$ we have $M' = M + N\mathbf{1}_t$ where $\mathbf{1}_t$ is the unit vector along the axis t; thus an invariant ι satisfies $\iota \cdot (N\mathbf{1}_t) = 0$ for all $\mathbf{1}_t$, and so is a solution of $\iota^{\mathrm{T}} N = 0$ (where ^{T} denotes transpose).

As will be seen, the notion of invariant plays a very important role in the applications of the tableau system to nets, for the easiest way to satisfy the inclusion condition on fix-point terminals is to find a suitable invariant. However, although methods for finding invariants are a fertile source of much and highly technical research, in our examples we shall only deal with invariants which may be seen 'by inspection', or constructed from the expected properties of the system and routinely verified.

Deadlock and liveness are of course fundamental properties of concurrent systems. In net theory their study has produced a rich body of theory concerning criteria for liveness etc. of certain classes of net. Some of these classes are both big enough to allow the modelling of interesting systems, and small enough to have nice criteria for properties such as liveness, and moreover have a rich structure theory. We now define some of the basic net-theoretic notions in this area, define some of the classes that will be referred to later, and give, by way of illustration, a sample of the results that are known (the Facts in this section can be found in [Rei85] or [Bes86] unless otherwise stated).

Definition 4.7. For a net $\mathcal{N}$ with initial marking M_0, a transition t is *live* iff $\forall M \in \overrightarrow{M_0} . \exists M' \in \overrightarrow{M} . M' \xrightarrow{t}$; that is, it is always possible for t to fire.

$(\mathcal{N}, M_0)$ is live iff t is live for all $t \in \mathbf{T}$. Note that this says every transition is live, not just that some transition is live, or that some transition can always fire.

$\mathcal{N}$ is *structurally* live iff $(\mathcal{N}, M_0)$ is live for some M_0.

A marking M of $\mathcal{N}$ is live iff $\forall t \in T . \exists M' \in \overrightarrow{M} . M' \xrightarrow{t}$; that is, each transition *may* fire (note that this definition says only that any transition may fire at least once, not that any transition is live). ◁

Hence we have that $(\mathcal{N}, M_0)$ is live iff M is live for all $M \in \overrightarrow{M_0}$.

So in mu-calculus terms, t is live if $M_0 \vDash \nu Y . [-]Y \wedge \mu Z . \langle - \rangle Z \vee \langle t \rangle \mathrm{tt}$.

Naturally, we also have

Definition 4.8. In a net N, a transition t is *M-dead* iff $\forall M' \in \overrightarrow{M}$. not $M' \xrightarrow{t}$; that is, t can never fire from M.

A marking M is dead iff not $M \longrightarrow$. ◁

Boundedness is an important property, though not one we shall have much occasion to use explicitly.

Definition 4.9. In a net $(\mathcal{N}, M_0)$ a place s is *bounded* iff $\exists n \in \mathbb{N} . \forall M \in \overrightarrow{M_0} . M(s) \leq n$.

$(\mathcal{N}, M_0)$ is bounded iff s is bounded for all $s \in \mathbf{S}$.

$\mathcal{N}$ is *structurally bounded* iff $(\mathcal{N}, M_0)$ is bounded for all M_0.

$(\mathcal{N}, M_0)$ is *safe* if $(\mathcal{N}, M_0)$ is bounded with a bound of 1 for every place.

$\triangleleft$

See [Rei85] for boundedness analysis.

The general class of nets for which special liveness criteria are studied is that of marked nets, in which all arcs have unit weight.

Definition 4.10. A net $\mathcal{N}$ is a *marked net* if range $\mathbf{F} \subseteq \{0, 1\}$, i.e. $\mathbf{F}$ is a true relation. $\triangleleft$

In net theory, the notions of liveness and deadlock are localized to subsets of places to give the ideas of a trap and a deadlock. A deadlock is a set of places which, once they lose all their tokens, will never be marked again; and dually a trap is a set of places which, once marked, never lose all their tokens. The following definitions ensure these properties.

Definition 4.11. Let $\mathcal{N}$ be a marked net and let $S \subseteq \mathbf{S}$.

S is a *deadlock* iff ${}^\bullet S \subseteq S^\bullet$.

S is a *trap* iff $S^\bullet \subseteq {}^\bullet S$. $\triangleleft$

Some elementary results about traps and deadlocks are

Fact 4.12. Let $\mathcal{N}$ be a marked net, M a marking, and $S \subseteq \mathbf{S}$.

 (i) The union of deadlocks (resp. traps) is a deadlock (resp. trap).

 (ii) If M is dead, $\{ s \mid M(s) = 0 \}$ is a non-empty unmarked deadlock of $\mathcal{N}$, and therefore

 (iii) For any M, if each non-empty deadlock of $\mathcal{N}$ contains a trap which is marked under M, then there is no dead marking in $\overrightarrow{M}$; that is, $(\mathcal{N}, M)$ does not deadlock. $\square$

An example of a theorem relating dynamic properties to structural properties (that is, properties of the net as a graph) is the following

Fact 4.13. (Thm 2.13 of [Bes86]) A net $\mathcal{N}$ is said to be strongly connected if $(\mathbf{S} \cup \mathbf{T}, \mathbf{F})$ is strongly connected when viewed as a directed graph. If $(\mathcal{N}, M_0)$ is live and safe, then $\mathcal{N}$ is strongly connected. $\square$

A particularly interesting class of marked nets is that of free-choice nets, where the interplay of concurrency and non-determinism is restricted, so that the following does not happen.

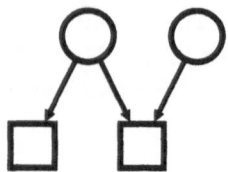

Definition 4.14. A marked net $\mathcal{N}$ is a *free-choice net* iff $\forall (s,t) \in \mathbf{F} \cap \mathbf{S} \times \mathbf{T}$. $s^\bullet = \{t\} \vee {}^\bullet t = \{s\}$. ◁

That is, no forward branched place may connect to a backward branched transition. The theory of free-choice nets is rich: see [Bes86] for an introduction. A good example is Commoner's criterion for liveness of a free-choice net:

Fact 4.15. [Com72], [Hac72] A free-choice net $(\mathcal{N}, M_0)$ is live iff every deadlock of $\mathcal{N}$ contains a trap that is marked under M_0. □

A still more restricted class is that of (structurally) live and bounded free-choice nets (LBFC nets), which have strong properties. For example, one property that has been studied in nets is that of having home states: this is a weak liveness property, being the existence of states that are always reachable.

Definition 4.16. Let $(\mathcal{N}, M_0)$ be a net with initial marking. A set $\mathbf{M}$ of markings is a *home space* iff $\forall M' \in \overrightarrow{M_0}. \exists M \in \mathbf{M}. M \in \overrightarrow{M'}$. A marking M is a *home state* if $\{M\}$ is a home space. ◁

Best and Voss showed the following nice result

Fact 4.17. [BeV84] If $\mathcal{N}$ is an LBFC net then $(\mathcal{N}, M_0)$ has a home state. □

In mu-calculus terms, M is a home state if $M_0 \vDash \nu Y.[-]Y \wedge \mu Z.\langle - \rangle Z \vee M$, and so this far from trivial theorem can be seen as the construction of a certain successful tableau!

Moreover, LBFC nets are especially interesting from the point of view of systems modelling, because recent work by Javier Esparza [Esp90] has shown that all LBFC nets may be produced from the elementary net

by the application of a small set of refinement rules. Not only is this interesting in itself, but it allowed him to prove an important theorem characterizing LBFC nets by the rank of the incidence matrix N (as defined after Definition 4.6).

4.1.3 Nets in systems modelling.

There is a plethora of approaches to building systems from nets. One possible top-level classification is into analytic and synthetic approaches: on the one hand, starting with a high-level model and refining it, and on the other, using rules to build complex systems from simpler nets. I am concerned mainly with synthesis, and here there are again two major divisions—the place-oriented and the transition-oriented. This division is not, of course, intended to be absolute, but I feel it distinguishes two different approaches to composition, which I hope will emerge in what follows.

A construction fundamental to all approaches is that which just lays two nets side by side.

Definition 4.18. Let $\mathcal{N}_1$ and $\mathcal{N}_2$ be two nets. Their *(disjoint) sum* $\mathcal{N}_1 + \mathcal{N}_2$ is the net $(\mathbf{S}_1 \cup \mathbf{S}_2, \mathbf{T}_1 \cup \mathbf{T}_2; \mathbf{F}_1 \cup \mathbf{F}_2)$. ◁

This corresponds to the parallel composition, with no synchronization, of the systems represented by the two nets. The 'place-oriented' constructions extend the sum by merging places in the two nets. The most general such technique is the notion of quotienting, introduced by Winskel in his work on categories of nets.

Definition 4.19. Let $\mathcal{N}$ be a net, $\mathbf{S}'$ a set, and $\beta : \mathbf{S} \to_\mu \mathbf{S}'$ a multirelation. The *quotient* $\mathcal{N}/\beta$ of $\mathcal{N}$ by β is $(\mathbf{S}', \mathbf{T}; \mathbf{F}')$, where $\mathbf{F}'(s', t) = \sum_{s \in S} \beta(s', s) \mathbf{F}(s, t)$ and similarly for $\mathbf{F}'(t, s)$. If $\mathcal{N}$ has an initial marking M_0, the initial marking M_0' of $\mathcal{N}'$ is given by $\mathbf{M}_0' = \sum_{s \in S} \beta(s', s) M_0(s)$. ◁

The quotient $\mathcal{N}/\beta$ is, then, constructed by replicating and merging places in $\mathcal{N}$, weighted according to the entries in β. A point to be aware of is that according to this definition, if we take the net

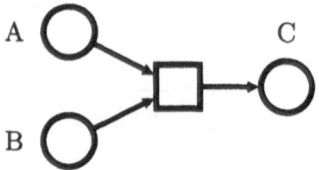

and quotient by the multirelation

$$\begin{array}{cc} & A' \quad C' \\ \begin{array}{c} A \\ B \\ C \end{array} & \left(\begin{array}{cc} 1 & 0 \\ 1 & 0 \\ 0 & 1 \end{array} \right) \end{array}$$

we get

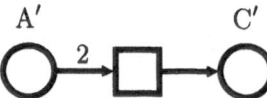

so that the class of marked nets is not closed even under quotienting by matrices over $\{0,1\}$. This is a consequence of taking multirelations and multisets as the basic entities; if one is considering safe nets, and viewing **F** as a relation, an appropriate notion of quotient is provided by replacing the summation by a set union. (See [Win84] for discussion of this.)

It is convenient to have a notation for the common operation of merging places.

Definition 4.20. Let $\mathcal{N}$ be a net, and let $S_1, \ldots, S_n$ be subsets of **S**. The net $\mathcal{N}/[s_1 = S_1, \ldots, s_n = S_n]$ is defined to be $\mathcal{N}/\beta$, where $\mathbf{S}' = (S - \bigcup_{i=1}^n S_i) \cup \{s_1, \ldots, s_n\}$ for $s_1, \ldots, s_n \notin \mathbf{S}$, and $\beta \colon \mathbf{S} \to_\mu \mathbf{S}'$ is given by $\beta(s', s) = 1$ iff $s = s' \lor \exists i . s' = s_i \land s \in S_i$, and $\beta(s', s) = 0$ otherwise. Thus each set S_i is collapsed to a single place s_i—note that the S_i are not necessarily disjoint. The new place name assignment $s_i =$ may be omitted if there is no need to name the merged place. It is sometimes convenient to allow $\beta(s', s)$ to take the value $1/k$ (for $k \in \mathsf{N}$) rather than 1 in order to remove a common factor from the weights of the arcs incident on s'; we write this by the notation $\mathcal{N}/[s = \frac{1}{k}S]$. ◁

The operation of merging two places into one has been employed in various ways. Berthelot in [Ber86] defines a notion of 'doubled place', in which he gives a criterion for two places s_1 and s_2 not to interfere with each other. Informally, this is so if no transition has both s_1 and s_2 as pre-places, and whenever some post-transition t_1 of s_1 is enabled (ignoring s_1) then s_2 is empty, and vice versa—that is, if s_1 has a potential role in enabling some transition, then s_2 must be completely empty. Under these conditions, s_1 and s_2 can be merged without changing the behaviour of the net.

Other place-oriented compositional techniques can be expressed in terms of sum and quotient. For example, in [SoM89] a notion of composition via shared places is defined which considers two nets $\mathcal{N}_1$ and $\mathcal{N}_2$ such that $\mathbf{T}_1 \cap \mathbf{T}_2 = \varnothing$ and $\mathbf{S}_1 \cap \mathbf{S}_2 = \mathbf{S}_c$ for some non-empty $\mathbf{S}_c$, and forms their composition just by taking the union of the two nets, i.e. $(\mathbf{S}_1 \cup \mathbf{S}_2, \mathbf{T}_1 \cup \mathbf{T}_2; \mathbf{F}_1 \cup \mathbf{F}_2)$. The sharing operation can be made explicit by using sum and quotient: let π be the natural projection from $\mathbf{S}_1 \uplus \mathbf{S}_2$ to $\mathbf{S}_1 \cup \mathbf{S}_2$; then the composition is $(\mathcal{N}_1 + \mathcal{N}_2)/\beta$ where $\beta(s', s)$ is 1 if $\pi(s) = s'$ and 0 otherwise.

A similar rule applies in the case of modelling systems such as reader–writer interlock, when one may design nets representing reader and writer processes, each with a copy of the resource, and then form a system by taking the sum of some number of reader and writer processes and then merging the resource places. We shall use the reader–writer system as a running example, so we now consider in more detail how the system might be built in a 'place-oriented' approach.

4.1.4 Building a reader–writer system by shared places.

Suppose that we wish to model the classic resource control problem where there is a resource which may be read or written, and there are many clients which wish to use the resource. The constraint on the system is that any process writing the resource must have exclusive access to it, so that no other process may be either reading or writing; but we wish to allow many processes to be reading simultaneously.

A natural way to represent such a system by means of Petri nets is the net in Figure 4.1. How may we consider this system to be composed of smaller nets? When following the shared places paradigm, it is natural to consider each reader or writer to be a process looking like Figure 4.2.

So each process comprises control-flow elements (the four leftmost elements) and a place representing the resource; the system is formed by merging each process's local resource place into one global resource place. However, to get the MRSW interlock, we must arrange for each writer's resource place to merge with every reader's resource place—and we may then merge the resulting places into one, to give the system of Figure 4.1. Alternatively, we may merge all the writers' resource places, and merge the

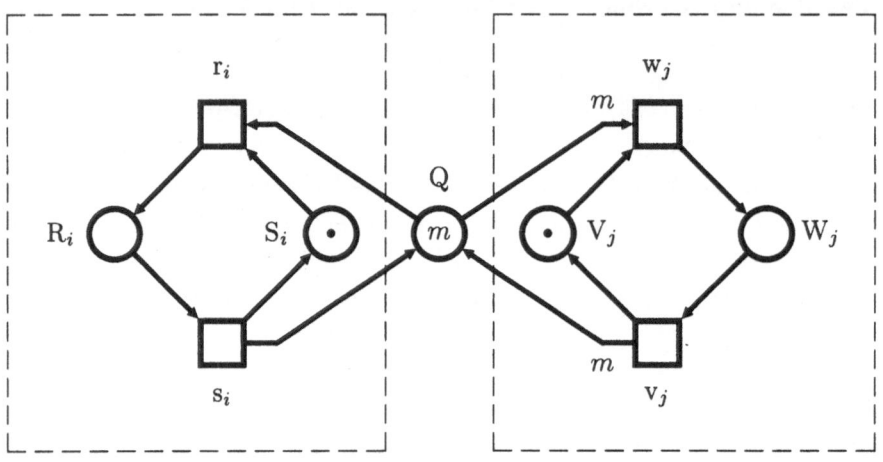

$$i = 1, \ldots, m \qquad\qquad j = 1, \ldots, n$$

Legend

For m reader processes   and n writer processes

S_i	process not reading	V_j	process not writing
r_i	process starts reading	w_j	process starts writing
R_i	process reading	W_j	process writing
s_i	process stops reading	v_j	process stops writing
Q	the resource		

Figure 4.1

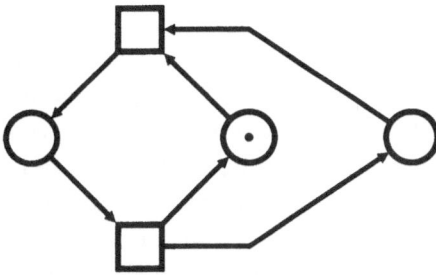

Figure 4.2

result with each reader's place, and then merge the result. If we name the processes $Reader_i$ and $Writer_j$, and name their respective resource places Q_{R_i} and Q_{W_j}, the two definitions of the system are written in our notation as

$$MRSW = \left(\left(\sum_{i=1}^{m} Reader_i + \sum_{j=1}^{n} Writer_j \right) \right.$$

$$/[Q_1 = \{Q_{W_1}, Q_{R_1}, \ldots, Q_{R_m}\}, \ldots, Q_n = \{Q_{W_n}, Q_{R_1}, \ldots, Q_{R_m}\}] \Big)$$

$$/[Q = \frac{1}{n}\{Q_1, \ldots, Q_n\}]$$

and

$$MRSW = \left(\sum_{i=1}^{m} Reader_i + \left(\sum_{j=1}^{n} Writer_j \right) / [Q_W = \{Q_{W_1}, \ldots, Q_{W_n}\}] \right)$$

$$/[Q_1 = \{Q_{R_1}, Q_W\}, \ldots, Q_m = \{Q_{R_m}, Q_W\}]/[Q = \{Q_1, \ldots, Q_m\}]$$

We have not yet specified the initial markings of the resource places—looking at the definition of quotient and the desired system, we see that we are compelled to assume that reader processes start with a marked resource place, and writers with an unmarked resource place; there is, for me at any rate, no obvious intuitive reason why this should be so. As a further warning against taking this example too seriously, note that it fails in the case that there are no reader processes; indeed, a consistent interpretation of the notation gives differing results in that case for the two definitions above: the first produces a place Q which is the merge of the writers' resource places, whereas the second calls this Q_W and also has an isolated place Q—and neither of these is how one naturally interprets Figure 4.1 with $m = 0$. In order to get a system that works for $m = 0$, it seems necessary to be a little less heavy-handed in conflating read permission and write permission—see the example in [Bra87]—but we need not pursue that here.

4.1.5 Transition-oriented design.

The foregoing should have reminded the reader of programming in a traditional imperative language; we turn now the transition-oriented approaches to net combinations, which have instead the flavour of CCS and CSP. This is presented here for its relevance for future work—while composition by synchronization has been studied in a trace-theoretic framework [Maz88], its utilization in compositional techniques for modal logic is difficult (see [Win90]) and requires further study.

The basic idea is exactly the same as that for places—take two nets, put them side by side, and merge transitions. So, if we liked, we could define the quotient of a net by a multirelation on the transition set by direct analogy with Definition 4.19.

Definition 4.21. Let $\mathcal{N}$ be a net, $\mathbf{T}'$ a set, and $\eta: \mathbf{T} \to_\mu \mathbf{T}'$ a multirelation. The quotient $\mathcal{N}/\eta$ is $(\mathbf{S}, \mathbf{T}'; \mathbf{F}')$ where $\mathbf{F}'(s, t') = \sum_{t \in \mathbf{T}} \eta(t', t)\mathbf{F}(s, t)$ and similarly for $\mathbf{F}'_{\cdot}(t', s)$.   ◁

However, this definition raises ontological questions about transitions—what does it mean to merge transitions, and in particular, what does it mean to 'multiply' a transition by having values greater than unity in η? It is natural to interpret the merge of t_1 and t_2 as their synchronization, but it is less clear what it would mean to merge a transition with itself.

At this point, it is worthwhile to digress on to the various categorical approaches to Petri nets, since they provide, in my view at least, mathematical justification for my opinion that merging transitions with themselves should not be allowed. The digression will be sketchy; the reader is referred to the original papers for details.

The approach we summarize now was started by Winskel in [Win84] and [Win85], and extended in [Win88].

Notation 4.22. If $\mathcal{N}$ is a net, $\phi\colon \mathbf{T} \to_\mu \mathbf{S}$ is the multirelation $\mathbf{F} \cap (\mathbf{S} \times \mathbf{T})$ (i.e. $\phi(t,s) = \mathbf{F}(s,t)$) and ψ is the converse of $\mathbf{F} \cap (\mathbf{T} \times \mathbf{S})$ (i.e. $\psi(t,s) = \mathbf{F}(t,s)$). ◁

Definition 4.23. The category **mNet** has as objects nets $\mathcal{N} = (\mathbf{S}, \mathbf{T}, \phi, \psi)$ and as morphisms pairs $(\beta, \eta)\colon \mathcal{N} \to \mathcal{N}'$ where $\beta\colon \mathbf{S} \to_\mu \mathbf{S}'$ and $\eta\colon \mathbf{T} \to_\mu \mathbf{T}'$ are such that $\beta\phi = \phi'\eta$ and $\beta\psi = \psi'\eta$. ◁

The commutativity requirements simply specify that the $^\bullet()$ and $()^\bullet$ relations are preserved (and therefore dynamic behaviour is preserved) by morphisms. Note that a quotient on places, as in Definition 4.19, corresponds to a morphism in which η is the identity.

Proposition 4.24. [Win85] The sum operation of Definition 4.18 is both the product and coproduct in **mNet**. □

Thus **mNet** is rather uninteresting. However, by imposing restrictions on the morphisms, notions of synchronization arise categorically.

Definition 4.25. **Net** is the category whose objects are nets and whose morphisms are pairs $(\beta, \eta)\colon \mathcal{N} \to \mathcal{N}'$ where $\beta\colon \mathbf{S} \to_\mu \mathbf{S}'$ is a multirelation and $\eta\colon \mathbf{T} \to_p \mathbf{T}'$ is a partial function. ◁

Note that this is quite a serious change, since composition of partial functions is not the same as composition of multirelations.

Proposition 4.26. [Win85] Let $\mathcal{N}_1$ and $\mathcal{N}_2$ be nets in **Net**. Their product $\mathcal{N} = \mathcal{N}_1 \times \mathcal{N}_2$ is $(\mathbf{S}_1 \uplus \mathbf{S}_2, \mathbf{T}_1 \uplus \mathbf{T}_2 \uplus (\mathbf{T}_1 \times \mathbf{T}_2), \phi, \psi)$ where for $t_i \in \mathbf{T}_i$ we have $\phi(t_i) = \phi_i(t_i)$ and $\phi(t_1, t_2) = \phi_1(t_1) + \phi_2(t_2)$, and similarly for ψ. □

So if we do not allow the 'replication' of transitions, the product is something that looks like a parallel composition with all synchronizations allowed (and also the unsynchronized events—if we further restrict η to be a total function, the unsynchronized events $\mathbf{T}_1 \uplus \mathbf{T}_2$ disappear from the product). Both these synchronizations also arise as products in categories defined in Meseguer and Montanari's view of nets as graphs with algebraic structure [MeM88], and there again the notion of morphism is such that replication of transitions is not allowed.

If we accept that this notion of product is a good method of combining nets, the question still remains of how to control the synchronization, since we do not usually want to retain all transitions, both synchronized and unsynchronized. The simplest approach is to take a leaf from the CCS book, and restrict away the unwanted events (by, for example, a suitable transition quotient). This can be make to sound less *ad hoc* by labelling transitions with elements of a synchronization algebra [Win88] (e.g. CCS actions) which determine which synchronizations occur: this labelling can be incorporated into the category, and then the product automatically contains only the desired transitions. Although the category theory is not directly relevant to us, the notions of synchronization algebra and products with respect to such are useful to have around, so we now define them.

Definition 4.27. A *synchronization algebra* is a set L of labels, not containing elements 0 and $\sharp$, with a binary, commutative and associative operation $\bullet$ on $L \cup \{0, \sharp\}$ satisfying

$$\forall \alpha, \alpha' \in L \cup \{0, \sharp\} . \, \alpha \bullet \sharp = \sharp \wedge (\alpha \bullet \alpha' = 0 \Rightarrow \alpha = \alpha' = 0).$$

Given nets $\mathcal{N}_1$ and $\mathcal{N}_2$ with their transitions labelled by functions $l_i \colon \mathbf{T}_i \to L$, let $\mathcal{N} = \mathcal{N}_1 \times \mathcal{N}_2$ be the product as in Proposition 4.26, and define $\hat{l} \colon \mathbf{T} \to L \cup \{\sharp\}$ by

$$\hat{l}(t_1, t_2) = l_1(t_1) \bullet l_2(t_2) \qquad \hat{l}(t_2) = 0 \bullet l_2(t_2) \qquad \hat{l}(t_1) = l_1(t_1) \bullet 0,$$

and let $T = \{\, t \in \mathbf{T} \mid \hat{l}(t) \neq \sharp \,\}$. Then we define the parallel composition of $(\mathcal{N}_1, l_1)$ and $(\mathcal{N}_2, l_2)$ with respect to L to be the net

$$\mathcal{N}_1 \times_L \mathcal{N}_2 = (\mathbf{S}, \mathbf{T}{\upharpoonright}T, \phi{\upharpoonright}T, \psi{\upharpoonright}T)$$

with labelling function $l = \hat{l}{\upharpoonright}T$.   ◁

The element 0 can be thought of as the label of a non-existent transition $*$; for convenience, the unsynchronized transitions t in the product are thought of as synchronizations with $*$. The element $\sharp$ is reserved to label transitions that should not appear in the final net.

4.2 Basic application to nets.

One question which we have not yet discussed at all is the problem of 'atomic propositions'—that is, what are the fundamental properties of nets from which we build more complex properties by means of the modal mu-calculus. This is to a large extent an empirical matter. For Petri nets, a very simple natural property is how many tokens are on a given place, and generalizing slightly, whether a marking satisfies some linear (in)equalities on the contents of places. These properties suffice for many examples.

Another question we have not considered is how we represent the sets that appear in tableau sequents—since they are in general infinite, we cannot simply enumerate their members. This is particularly important when considering implementations, since some method of finitely presenting sets must be built in to a tool. Again, this turns out to be an empirical question, since with a little ingenuity quite simple nets and formulae can produce extremely complex sets. However, in practice things are not so bad; linear inequalities again suffice for 'sensible' properties of 'sensible' systems. Therefore we shall take linear equalities as standard, incorporating them into the set of variables in the following manner.

Definition 4.28. Let $\mathcal{N} = (\mathbf{S}, \mathbf{T}; \mathbf{F})$ be a net. The variable set Var of the modal mu-calculus is deemed to contain variable symbols of the form $a_1 s_1 + \cdots + a_n s_n = b$ and $a_1 s_1 + \cdots + a_n s_n \leq b$ where $n \geq 1$, $b \in \mathbf{Z}$ and $s_i \in \mathbf{S}$ and $a_i \in \mathbf{Z}$ for $1 \leq i \leq n$. These variable symbols are referred to as *atomic propositions*. A valuation $\mathcal{V}$ is *standard* for $\mathcal{N}$ if it assigns values to these atomic propositions in the natural way, that is,

$$\mathcal{V}(a_1 s_1 + \cdots + a_n s_n = b) = \{\, M \mid \sum_{i=1}^{n} a_i M(s_i) = b \,\}$$

and similarly for $\leq$.   ◁

Henceforth, standard valuations are assumed.

Notation 4.29. Given a net $\mathcal{N}$ and a predicate Φ on markings in some language, $\{\Phi\}$ denotes the set of markings satisfying Φ. Henceforth the language is assumed to be boolean combinations of atomic propositions. ◁

Notation 4.30. Henceforth the definition lists Δ will be omitted from tableaux, since they are determined entirely by the formulae and rules of the tableau.   ◁

4.2.1 A simple example.

The first example is a very simple net, but a complex formula, namely a 'finitely often' formula. Consider the net of Figure 4.3.

The behaviour of this net is just that a fires, adding tokens to B, until b fires; and then c fires until B is again depleted. So we should be able to prove that c fires only finitely often, that is,

$$\Phi \stackrel{\text{def}}{=} \mu Y . \nu Z . [c] Y \wedge [-c] Z.$$

This formula should be satisfied by the initial marking, but since this is an infinite system we shall need to use Thin. Either by insight or by trying to build a tableau without Thin, it seems that the set $\{A + C = 1\}$ is promising. So a suitable candidate for a proof is the tableau of Figure 4.4.

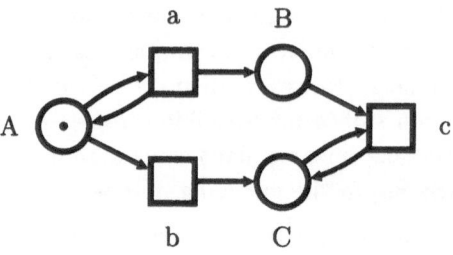

Figure 4.3

$$\frac{\{\,A = 1 \wedge B = 0 \wedge C = 0\,\} \vdash_{()} \mu Y.\nu Z.[c]Y \wedge [-c]Z}{\{\,A + C = 1\,\} \vdash_{()} \mu Y.\nu Z.[c]Y \wedge [-c]Z}$$

$$\frac{{}_1\{\,A + C = 1\,\} \vdash_\Delta U}{\{\,A + C = 1\,\} \vdash_\Delta \nu Z.[c]U \wedge [-c]Z}$$

$$\frac{{}_2\{\,A + C = 1\,\} \vdash_{\Delta'} V}{\{\,A + C = 1\,\} \vdash_{\Delta'} [c]U \wedge [-c]V}$$

$$\frac{{}_3\{\,A + C = 1\,\} \vdash_{\Delta'} [c]U \qquad\qquad {}_5\{\,A + C = 1\,\} \vdash_{\Delta'} [-c]V}{{}_4\{\,A = 0 \wedge C = 1\,\} \vdash_{\Delta'} U \qquad\qquad {}_6\{\,\Phi'\,\} \vdash_{\Delta'} V}$$

Figure 4.4

The non-trivial steps are at nodes 3 and 5—but it is easy to see that the set of c-successors of $\{\,A + C = 1\,\}$ is $\{\,A = 0 \wedge C = 1\,\}$, and likewise for $-c$-successors. (At node 6, Φ' is really $(A + C = 1) \wedge (B = 0 \Rightarrow C = 1)$, but all that matters for the success of the tableau is that it should imply $A + C = 1$; there is no need to calculate it.)

To prove soundness, we must show well-foundedness of $\sqsubset_1$. Examination of the tableau shows us that $M \sqsupset_1 M'$ iff $M \xrightarrow{-c}{}^* \circ \xrightarrow{c} M'$, and so it suffices to find some non-negative measure which strictly decreases under $\xrightarrow{-c}{}^* \circ \xrightarrow{c}$. On the set $\{\,A + C = 1\,\}$ such a measure δ is provided by $\delta(M) = (M(A), M(B))$ with the lexicographic ordering: if $M \xrightarrow{-c}{}^* M'' \xrightarrow{c} M'$ then $M''(C) = 1$ and $M'(A) = 0$; if the $(-c)^*$ sequence contains the event b then $M(A) = 1$ so $\delta(M) > \delta(M')$, and otherwise $M(A) = M''(A) = 0$ so the sequence is empty, and then $M'(B) = M(B)-1$, so again $\delta(M) > \delta(M')$.

This very simple example illustrates already the appearance of net invariants in the use of the tableau system: if we use invariants as descriptions of the states labelling fix-points, we are sure to get σ-terminals satisfying the inclusion condition. It is worth noting that $\{\,A + C = 1\,\}$ is exactly those markings satisfying Φ that are reachable from the given initial marking; in the absence of a given initial marking, $\|\Phi\|$ is in fact $\{\,(A + C = 1) \vee (A = 0)\,\}$. This example illustrates how much easier the tableau system is to use than calculating approximants: the reader who doubts this is invited to try calculating the approximants in this case—the task requires keeping a very clear head for the book-keeping even in so simple a net as this. (Solution: $\|\Phi^n\| = \{\,A = 0 \wedge (B < n \vee C = 0)\,\}$ and $\|\Phi\| = \|\Phi^{\omega+1}\| = \|\Phi^\omega\| \cup \{\,A = 1 \wedge C = 0\,\}$.)

4.2.2 A slot machine.

The second example was first presented in [Bra91], and demonstrates the proof of safety and liveness properties on a less trivial net.

Consider the net of Figure 4.5. This represents a simple slot machine which accepts a coin, and then pays out a random amount between zero and the amount in its bank; and a gambler who has an inexhaustible supply of credit—we shall assume that the creditor, though infinitely rich, is irrational enough to demand his money back eventually. Since the machine starts with an empty bank, the gambler is always in debt; we should like to encourage her to continue to play by proving to her that it is always possible for her to recover her losses, that is, it is always possible that $C - W = 0$, which in the mu-calculus is

$$\nu Y.(\mu Z.(C - W = 0) \vee \langle\rangle Z) \wedge [\,]Y.$$

As always when using the tableau system, a crucial step is choosing the right set of states for the root sequent of the tableau, or rather the right set to weaken the root to. We shall mention later a couple of other possibilities for deciding how to do this, but for the moment let us continue to use invariants and inspiration. By looking at the design of the net, it is apparent that we ought to have $P + R = 1$, since these places are essentially a finite state control for the machine. We can also see at once that $M = 1$ remains true. Finally, it ought to be the case that $C = S + B + W$, since the left-hand side is a note of the total amount of money put into the system by the player and the right-hand side is that amount. So the set of sensible markings for which we may hope to prove the property is

$$\{\,\Phi \overset{\text{def}}{=} (M = 1) \wedge (P + R = 1) \wedge (C = S + B + W)\,\}.$$

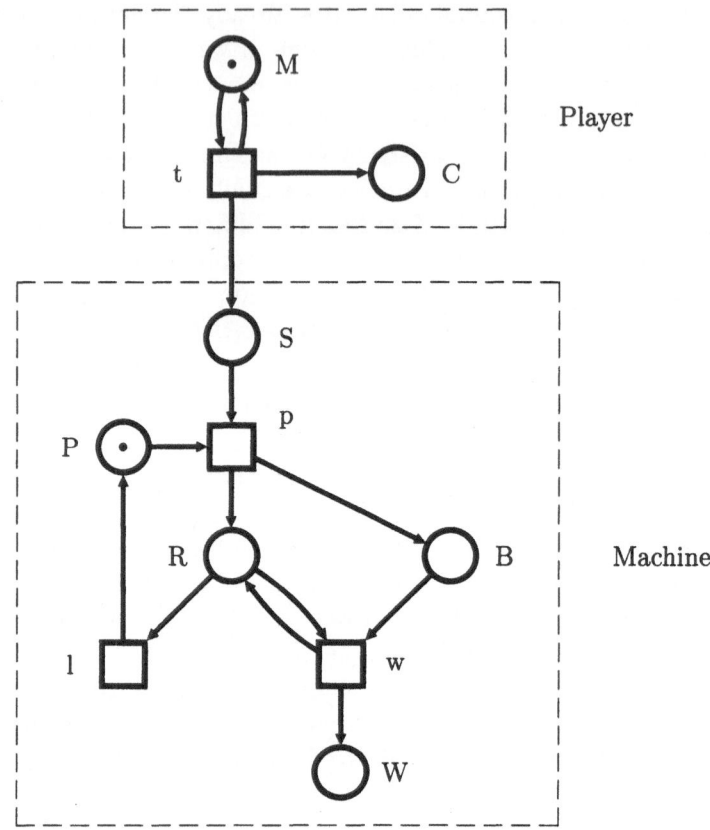

Player

Machine

Legend

M	the player's source of credit
t	the player puts a coin in the slot
C	how many coins the player has spent
S	the slot (with unbounded storage capacity)
p	the PLAY button
P	the machine is ready to play
B	the machine's bank
R	the reels are spinning
W	the player's winnings
l	lose: the machine stops paying out and is ready to play again
w	win: the machine pays out a coin, and keeps spinning the reels

Figure 4.5

A slot machine.

The tableau for this formula and set of markings is set out in Figure 4.6.

$$\{\Phi\} \vdash \nu Y.(\mu Z.(C = W) \vee \langle\rangle Z) \wedge []Y$$

$$\overline{\{\Phi\} \vdash U}$$

$$\overline{\{\Phi\} \vdash (\mu Z.(C = W) \vee \langle\rangle Z) \wedge []U}$$

$$\frac{\{\Phi\} \vdash \mu Z.(C = W) \vee \langle\rangle Z \qquad {}_5\{\Phi\} \vdash []U}{{}_1\{\Phi\} \vdash V \qquad\qquad {}_6\{\Phi''\} \vdash U}$$

$${}_2\{\Phi\} \vdash (C = W) \vee \langle\rangle V$$

$$\{\quad \Phi \wedge (S + B > 0) \\ \wedge \wedge (B = 0)\} \vdash (C = W)$$

$${}_3\{\Phi \wedge (S + B > 0)\} \vdash \langle\rangle V$$

$$\{\quad (\Phi \wedge (S + (B + 1) > 0) \wedge (B \geq 0)) \\ \vee (\Phi \wedge (S + B > 0) \wedge (B = 0)) \\ \vee (\Phi \wedge ((S + 1) + (B - 1) > 0) \\ \wedge (B - 1 = 0) \wedge (S \geq 0)) \\ \vee (\Phi \wedge ((S - 1) + B > 0)) \\ {}_4 \quad \} \vdash V$$

Figure 4.6

Some notes on the rules:

2 At this disjunction, we follow our recommendation and make the two consequent sets disjoint. It is usually advantageous to put as many states as possible into a branch with no variables (since this contributes to the first unfolding of least fix-points), so we put into the left branch exactly those markings that satisfy $C = W$.

3 This is the most complex part of the tableau. Following the construction in the completeness proof, we wish to choose here the successor that is in the lowest unfolding. Intuitively, this corresponds to choosing a successor that gets us nearer the goal of $C - W = 0$, which we may do by making the machine lose if possible. To achieve this, we partition the markings of the premise into the following four disjoint subsets, where $\Phi' \stackrel{\text{def}}{=} \Phi \wedge (S + B) > 0$:

  (i) $\{\Phi' \wedge (R = 1) \wedge (B > 0)\}$

  (ii) $\{\Phi' \wedge (R = 1) \wedge (B = 0)\}$

  (iii) $\{\Phi' \wedge (P = 1) \wedge (S > 0)\}$

  (iv) $\{\Phi' \wedge (P = 1) \wedge (S = 0)\}$

and for all states in class (i) we fire w, in (ii) l, in (iii) p, and in (iv) t; this gives the four disjuncts in the conclusion.

4 This is a successful mu-terminal: every disjunct implies Φ, so the node satisfies the inclusion condition; and $\sqsubset_1$ is well-founded, for if $M@1 \longrightarrow M'@4$, the path must be $M@1$, $M@2$, $M@3$, $M'@4$, so we need only consider M satisfying Φ'. Then we observe

(a) if M is in class (i), then either B > 1, in which case M' is in class (i) with strictly smaller B, or B = 1, in which case M' is in class (ii) ; and in either case, S is unchanged;

(b) if M is in class (ii), then M' is in class (iii) (because B = 0 $\Rightarrow$ S > 0 since S + B > 0); and S is unchanged;

(c) if M is in class (iii), then M' is in class (i) and S decreases by one;

(d) if M is in class (iv), then either M' has S = B = 0, in which case $M' \not\sqsupset_1$, or M' is in class (iii).

If we now impose an order $\prec$ on markings by the lexicographic ordering on the pair (S, B), we see from the above that if $M \sqsupset_1 M' \sqsupset_1 M''$ then $M'' \prec M$; and since $\prec$ is well-founded, so is $\sqsubset_1$.

5 (Box) Here we must consider all possible firings of transitions for each marking. But since all we want is that 6 should satisfy the inclusion condition, it suffices to note that Φ remains true under any firing, and so

6 is a successful ν-terminal, since Φ'', whatever it be, implies Φ.

4.3 Using schematic tableaux.

The above examples show how the tableau system can be used to prove interesting properties of some simple systems modelled as nets. It is noticeable that to do the proof (at least as presented above) we used our understanding of the system being modelled, and of the model, to guide the tableau construction. In general, this is a feature, not a flaw, since we know that complete automation is not possible, and we expect a designer of a system to understand it well enough to have an idea as to why true properties work. However, it is useful to have less *ad hoc* techniques to apply when inspiration fails, and ways to deal with complex systems that do not treat the systems as monoliths. Ideally, we would have a compositional calculus of nets which would provide us with a set of compositional tableau-building techniques. However, the problem of compositionality for nets is still wide open, with many competing approaches none of which is obviously 'right', and none of which provides, as yet, a compositional system for the modal mu-calculus. Therefore we shall concentrate on techniques rather than rules, and heuristics rather than algorithms.

The first technique we shall consider is the obvious but nonetheless important one of parametrizing tableaux. This is not so much a replacement for insight as a means of using insight about the nature of a system with a number of components: given a system with a variable number of similar components, we construct a tableau schema which provides proofs for the system with any number of components.

4.3.1 A simple parametrized safety proof.

To demonstrate this technique, we consider the reader–writer system and the proof of a safety property. There are several different ways of formulating the safety properties, with different emphases on places and transitions. For the moment, consider only half the safety property, the half that says 'if some process is reading, no process may be writing'. A direct translation of this is

$$\forall \mathbf{G}\left(\left(\bigvee_{1 \leq i \leq m} (R_i = 1)\right) \Rightarrow \bigwedge_{1 \leq j \leq n} (W_j = 0)\right)$$

which with a little arithmetic becomes

$$\forall \mathbf{G}\left(\sum_i R_i > 0 \Rightarrow \sum_j W_j = 0\right)$$

put into the mu-calculus as

$$\nu Z.\left(\sum_i R_i > 0 \Rightarrow \sum_j W_j = 0\right) \wedge [-]Z.$$

This is a very static expression of the safety property: it assumes that what matters is the state, as determined by the number of tokens in each place. We could instead take a more transition-based view, and say that the safety property is actually 'once any process starts reading, no process may start writing until the first process stops reading'. This can be written as

$$\nu Y. \bigwedge_{1 \leq i \leq m} \left([r_i]\nu Z_i.[\{w_j \mid 1 \leq j \leq n\}]\mathrm{ff} \wedge [-s_i]Z_i\right) \wedge [-]Y$$

which is considerably more complex—this is only to be expected, since it makes no use at all of the fact that the underlying transition system is a Petri net, whereas the first formula implicitly uses the token game rules to translate conditions on events into conditions on states.

To prove this property, in its first form, say, we present a tableau schema with nodes parametrized by the number of processes, such that every instantiation of the the schema is a successful tableau. It is convenient to extend the modal mu-calculus to allow finite conjunction and disjunction, and add the obvious schematic tableau rules:

$$\frac{S \vdash_\Delta \bigwedge_{i \in I} \Phi_i}{S \vdash_\Delta \Phi_i}$$

and

$$\frac{S \vdash_\Delta \bigvee_{i \in I} \Phi_i}{S_i \vdash_\Delta \Phi_i}$$

where $\bigcup_{i \in I} S_i = S$, which when instantiated give a $|I|$-way branch in the tableau.

Now consider a tableau for the first formula, which in positive normal form is

$$\nu Z.\Big(\sum_{1 \le j \le n} W_j = 0 \vee \sum_{1 \le i \le m} R_i = 0 \Big) \wedge [-]Z.$$

We again use our insight about the construction of the system to select some invariants that should be relevant, in particular $(R_i + S_i)$ and $(V_j + W_j)$ (again, this is the finite state control of the separate processes), and the crucial invariant $Q + \sum_i R_i + m \sum_j W_j)$ (here and henceforth we assume for the sake of brevity that i and j range over $1, \ldots, m$ and $1, \ldots, n$). Let

$$\Phi \stackrel{\text{def}}{=} \bigwedge_i (R_i + S_i = 1) \wedge \bigwedge_j (V_j + W_j = 1) \wedge \Big(Q + \sum_i R_i + m \sum_j W_j = m \Big).$$

A tableau schema for this is very simple:

$$\frac{\{\Phi\} \vdash \nu Z.(\sum_j W_j = 0 \vee \sum_i R_i = 0) \wedge [-]Z}{\{\Phi\} \vdash U}$$

$$\frac{\{\Phi\} \vdash (\sum_j W_j = 0 \vee \sum_i R_i = 0) \wedge [-]U}{\frac{\{\Phi\} \vdash \sum_j W_j = 0 \vee \sum_i R_i = 0 \qquad\qquad \{\Phi\} \vdash [-]U}{\{\Phi \wedge \sum_j W_j = 0\} \vdash \sum_j W_j = 0 \quad \{\Phi \wedge \sum_i R_i = 0\} \qquad \{\Phi\} \vdash U}}$$
$$\vdash \sum_i R_i = 0$$

The work here is in checking that the disjunction rule is correct, i.e. that $\{\Phi \wedge \sum_j W_j = 0\} \cup \{\Phi \wedge \sum_i R_i = 0\} = \{\Phi\}$, which is just a matter of manipulating equations.

In this case, we have used so much knowledge of invariants that the tableau is all but trivial; if we try instead to prove the second formulation of the property, more of this concealed knowledge appears in the tableau, in Figure 4.7 (where $\boldsymbol{w} \overset{\text{def}}{=} \{ w_j \mid 1 \leq j \leq n \}$).

$$
\begin{array}{c}
\dfrac{\{\Phi\} \vdash \nu Y. \bigwedge_i \big([r_i]\nu Z_i.[\boldsymbol{w}]\mathrm{ff} \wedge [-s_i]Z_i\big) \wedge [-]Y}{\{\Phi\} \vdash U}
\end{array}
$$

$$
\dfrac{\{\Phi\} \vdash \bigwedge_i \big([r_i]\nu Z_i.[\boldsymbol{w}]\mathrm{ff} \wedge [-s_i]Z_i\big) \wedge [-]U}
$$

$$
\begin{array}{cc}
\{\Phi\} \vdash \bigwedge_i \big([r_i]\nu Z_i.[\boldsymbol{w}]\mathrm{ff} \wedge [-s_i]Z_i\big) & \{\Phi\} \vdash [-]U \\[4pt]
\hline
{}_1\{\Phi\} \vdash [r_i]\nu Z_i.[\boldsymbol{w}]\mathrm{ff} \wedge [-s_i]Z_i & \{\Phi\} \vdash U
\end{array}
$$

$$
\dfrac{\{\Phi \wedge S_i = 0\} \vdash \nu Z_i.[\boldsymbol{w}]\mathrm{ff} \wedge [-s_i]Z_i}{\{\Phi \wedge S_i = 0\} \vdash V_i}
$$

$$
\{\Phi \wedge S_i = 0\} \vdash [\boldsymbol{w}]\mathrm{ff} \wedge [-s_i]V_i
$$

$$
\begin{array}{cc}
\dfrac{\{\Phi \wedge S_i = 0\} \vdash [\boldsymbol{w}]\mathrm{ff}}{\{\,\} \vdash \mathrm{ff}} & \dfrac{{}_2\{\Phi \wedge S_i = 0\} \vdash [-s_i]V_i}{\{\Phi \wedge S_i = 0\} \vdash V_i}
\end{array}
$$

Figure 4.7

In this tableau, at 1 and 2 we have to work out the result of firing a transition from a marking in the premise: at 1, if $M \in \{\Phi\}$ and $M \xrightarrow{r_i} M'$, then by plugging the firing rule into Φ, we get that M' satisfies

$$
\bigwedge_{k \neq i} (R_k + S_k = 1)
$$

$$
\wedge \, ((R_i - 1) + (S_i + 1) = 1)
$$

$$
\wedge \, \bigwedge_j (V_j + W_j = 1)
$$

$$
\wedge \left((Q+1) + \sum_{k \neq i} R_k + (R_i - 1) + m \sum_j W_j = m \right) .
$$

which boils down to $\Phi \wedge (S_i = 0)$, since we also have the implicit constraint that the number of tokens is non-negative. Now at 2 we have to show that this latter formula ensures that no $\boldsymbol{w}$ event may fire, and this follows since the third clause in Φ together with the information that $R_i = 1$ gives $Q < m$.

4.3.2 A parametrized liveness proof.

Parametrized tableaux can of course be built for more interesting properties than simple safety properties. We now give an example of such a proof, to show how the well-foundedness proof is parametrized.

Let us consider a desirable liveness property for the reader–writer system, namely that if a writer wants to write, then eventually it does write. The system we have used so far does not have this property, so if we hope to prove it we must have a better designed system. Achieving full and fair liveness would make an already somewhat unwieldy example uncomfortably large, so we shall consider the restricted liveness requirement that the reader processes should not be able indefinitely to defer writing by never letting go of the resource completely. A system which addresses this requirement was used as an example in [Bra91] and [BrS91], and is given in Figure 4.8.

The system works by having an additional interlock place P; when the token is removed from this place by a writer process firing a u ('request to write') event, no further read actions (or write-request actions) may occur, and so eventually all the currently reading processes stop reading, and the writer can take control of the resource.

A formulation of the liveness property (compromising this time between transition-based and place-based formulations) is

$$\nu Y.[-]Y \wedge \bigwedge_j [u_j](\mu Z.(W_j = 1) \vee [-]Z).$$

Following the previous example the basic invariant is

$$\Phi \stackrel{\text{def}}{=} \bigwedge_j (U_j + V_j + W_j = 1)$$

$$\wedge \bigwedge_i (R_i + S_i = 1)$$

$$\wedge \; (Q + m\sum_j W_j + \sum_i R_i = m)$$

$$\wedge \; (P + \sum_j U_j = 1)$$

and the tableau is now a little more complex: let $\Psi_j \stackrel{\text{def}}{=} (U_j = 1) \wedge (P = 0)$ and see Figure 4.9.

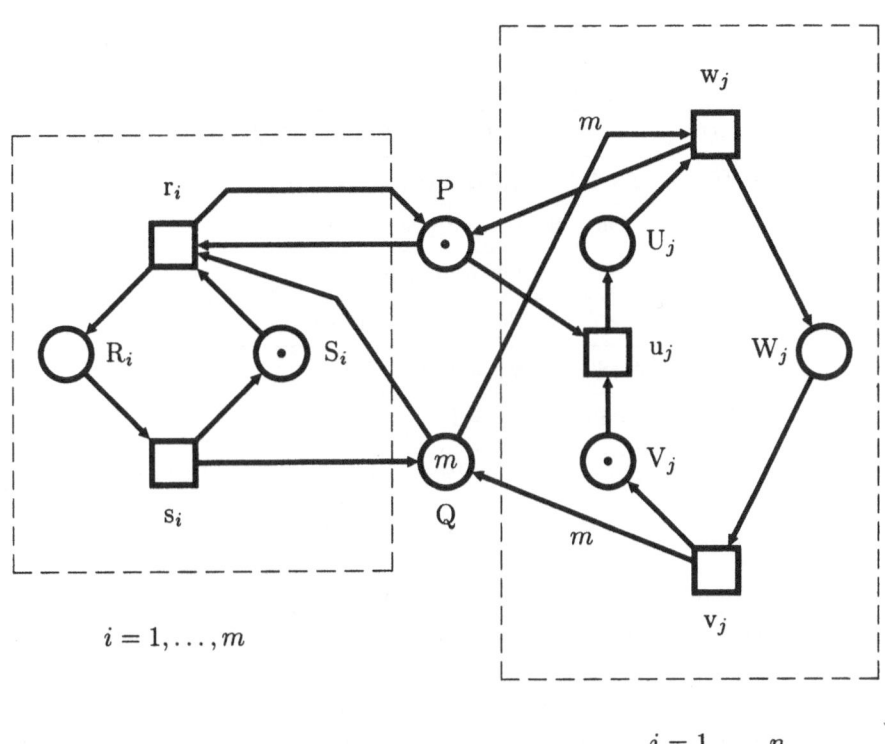

$$i = 1, \ldots, m$$

$$j = 1, \ldots, n$$

Legend

	For m reader processes		and n writer processes
S_i	process not reading	V_j	process not writing
r_i	process starts reading	w_j	process starts writing
R_i	process reading	W_j	process writing
s_i	process stops reading	v_j	process stops writing
		u_j	process requests write access
		U_j	process waiting to write
Q	the resource		
P	an interlock		

Figure 4.8

$$\frac{\{\,\Phi\,\} \vdash \nu Y.[-]Y \wedge \bigwedge_j [u_j](\mu Z.(W_j = 1) \vee [-]Z)}{\{\,\Phi\,\} \vdash U}$$

$$\{\,\Phi\,\} \vdash [-]U \wedge \bigwedge_j [u_j](\mu Z.(W_j = 1) \vee [-]Z)$$

$$\frac{\{\,\Phi\,\} \vdash [-]U \qquad \{\,\Phi\,\} \vdash \bigwedge_j [u_j](\mu Z.(W_j = 1) \vee [-]Z)}{}$$

$${}_1\{\,\Phi'\,\} \vdash U \qquad {}_2\{\,\Phi\,\} \vdash [u_j]\mu Z.(W_j = 1) \vee [-]Z$$

$$\{\,\Phi \wedge \Psi_j\,\} \vdash \mu Z.(W_j = 1) \vee [-]Z$$

$${}_3\{\,\Phi \wedge \Psi_j\,\} \vdash V_j$$

$${}_4\{\,\Phi \wedge (\Psi_j \vee (W_j = 1))\,\} \vdash V_j$$

$${}_5\{\,\Phi \wedge (\Psi_j \vee (W_j = 1))\,\} \vdash (W_j = 1) \vee [-]V_j$$

$${}_6\{\,\Phi \wedge (W_j = 1)\,\} \vdash W_j = 1 \qquad {}_7\{\,\Phi \wedge \Psi_j\,\} \vdash [-]V_j$$

$${}_8\{\,\Phi \wedge (\Psi_j \vee (W_j = 1))\,\} \vdash V_j$$

Figure 4.9

As before, we have to do some rather tedious calculation: at node 1, to get the success of this ν-terminal all we need to do is check that Φ' implies Φ, and this follows from the invariance of Φ. At 2, we have to calculate the result of firing u_j at a marking satisfying Φ, and we get $\Phi \wedge \Psi_j$. The use of Thin at 3 is due to the fact that at node 7 it may be that w_j fires, in which case Ψ_j becomes false, so the inclusion condition for 8 would fail if we did not thin at 3. Then at 4 we adopt the usual policy of putting as much as possible into the variable-free branch of the disjunction, and putting the rest in the other branch (Ψ and $W_j = 0$ are mutually exclusive if Φ is true, so the sets labelling the disjuncts are indeed disjoint, as in our recommendation). Now we must check that 8 satisfies the inclusion condition, which it does. (By calculation, when an event fires from a marking satisfying $\Phi \wedge \Psi_j$, either Ψ_j remains true, or $W_j = 1$ becomes true; this is how we find that we must thin at 3. This approach of building a tableau and then finding that Thin must be used and going back to an earlier point, is an alternative to using repeated unfolding. In the infinite case, some intelligence may be required to recognize when an infinite sequence of such operations is being performed, so that one may thin to the necessary set in one swoop; whether it is easier to do this by repeatedly retrying the tableau from a certain point, or whether by repeated unfolding until a pattern emerges, depends on the taste of the user.)

Finally, the less routine task of checking well-foundedness of $\sqsubset_3$ (remembering that nodes 3–8 should really be indexed by j, so there are n relations). As before, we have an intuitive idea of what this relation means: $M \sqsupset_3 M'$ means that M' is closer than M to the 'goal' of the fix-point, a state in which $W_j = 1$, a preliminary of which is that w_j should fire. So an ordering to prove well-foundedness should be a measure of how far away from firing w_j is. The simplest such measure is just to count the shortfall of tokens in the resource place: $m - Q$, or, using the invariant Φ, $\sum_i R_i + m \sum_{1 \le k \le n} W_k$. Let $\delta(M) = \sum_i M(R_i) + m \sum_k M(W_k)$. We now show that $M \sqsupset_3 M' \Rightarrow \delta(M) > \delta(M')$, unless $\nexists M'' . M' \sqsupset_3 M''$ (in which case well-foundedness is true anyway). So suppose $M \sqsupset_3 M' \sqsupset_3 M''$. Then both M and M' are in the set labelling node 7, so for both we have $P = 0$ and $U_j = 1$. So the only transitions that can fire at M to give M' are s_i for any i and v_k for any k, and in either case $\delta(M) > \delta(M')$. So our tableau is successful.

4.4 Using limited reachability analysis—the coverability graph.

If we do not have sufficient insight into a net to 'construct' a tableau at one swoop, it will be necessary to do some exploration of the state space. One way to do this is to build a tableau starting from the markings of interest and unfolding repeatedly until either a successful tableau is produced or enough of a pattern emerges to suggest an appropriate use of Thin. However, an alternative is to use established techniques from net theory to perform limited reachability analysis and try to build a tableau from the result. One such technique is the *coverability graph*.

The coverability graph [Rei85] of a net $(\mathcal{N}, M_0)$ is a finite graph such that every reachable marking M of $\mathcal{N}$ is 'covered by' a node of the graph, in the sense that either M appears in the graph, or M is part of an infinite sequence of strictly increasing markings, and there is a node in the graph representing this entire sequence. The main use of coverability graphs in net theory is to prove certain boundedness and liveness properties: since coverability graphs are finite, this gives decision procedures for those properties.

The nodes of a coverability graph are just markings with added provision that a place may be marked by ω, representing an increasing firing sequence of markings with unbounded value on that place. The reason for requiring the sequence to be increasing is that then any transition enabled

at a given point in the sequence must also be enabled at all later points in
the sequence, so allowing the proof of the coverability properties.

The definition of coverability graph below is from [Rei86], modified to
produce a graph rather than a tree.

Definition 4.31. Let $\mathcal{N} = (\mathbf{S}, \mathbf{T}; \mathbf{F})$ be a net with initial marking M_0.
Define a sequence of labelled directed graphs $G_0, G_1, \ldots$ with vertices N in
$(\mathbb{N} \cup \{\omega\})^{\mathbf{S}}$ and edges (N, t, N') labelled by transitions thus:

- $G_0 = (\{M_0\}, \varnothing)$
- if $G_i = (V_i, E_i)$, then construct G_{i+1} so: if there exist $N \in V_i$ and
  $t \in \mathbf{T}$ such that
    (i) t is enabled at N (where the token game is extended to infinite
        markings by taking $\forall n \in \mathbb{N} . \, \omega + n = \omega - n = \omega$) and
    (ii) $\nexists N' . \, (N, t, N') \in E_i$
  let $N' = N - {}^{\bullet}t + t^{\bullet}$; let P be the set of nodes N'' in G_i such that
  $N'' \leq N'$ and there is a path from N'' to N' in G_i extended by N' and
  (N, t, N'); then define $\hat{N}$ by
    - $\hat{N}(s) = \omega$ if there exists $N'' \in P$ such that $N''(s) < N'(s)$,
    - $\hat{N}(s) = N'(s)$ otherwise
  and let $G_{i+1} = (V_i \cup \{\hat{N}\}, E_i \cup \{(N, t, \hat{N})\})$; if no such N, t exist, then
  let $G_{i+1} = G_i$.

The coverability graph of $(\mathcal{N}, M_0)$ is $\bigcup_{i=0}^{\infty} G_i$.   ◁

The motivating properties of coverability graphs are given in the following theorem, the proof of which may be found in [Rei86] or [Rei85].

Fact 4.32. Let G be the coverability graph of a net $(\mathcal{N}, M_0)$.

(i) For every firing sequence $M_0 \xrightarrow{t_1} \cdots \xrightarrow{t_n} M_n$ there exists a path
    $N_0 t_1 N_1 \ldots N_{n-1} t_n N_n$ in G such that $N_0 = M_0$ and $\forall 1 \leq i \leq n . \, M_i \leq$
    N_i; that is, every reachable marking of $\mathcal{N}$ is covered by a node of G.

(ii) For a node N of G, a covering set $\mathbf{M}_N \subseteq \overrightarrow{M_0}$ of reachable markings is
    a minimal set such that

$$\forall i \in \mathbb{N} . \, \exists M \in \mathbf{M}_N . \, \forall s \in \mathbf{S} .$$

$$(N(s) < \omega \Rightarrow M(s) = N(s)) \wedge (N(s) = \omega \Rightarrow M(s) \geq i)$$

that is, a set such that the ω-places of $\mathcal{N}$ are (simultaneously) unbounded in $\mathbf{M}_N$. Every node of G has a covering set.

(iii) If $\mathcal{N}$ is finite so is G.   □

Thus an unbounded node in a coverability graph has an outgoing t arc
if almost all markings covered by that node enable t. This of course means

that termination properties are not preserved in the coverability graph, as we see in the following example.

Recall the example net of subsection 4.2.1. Following the algorithm, the coverability graph is given in Figure 4.10, where the small numbers show at which stage each node and arc was added, and abc denotes the node N with $N(A) = a$, $N(B) = b$ and $N(C) = c$.

Having constructed the coverability graph, we can now attempt to construct a tableau (for our original formula $\mu Y.\nu Z.[c]Y \wedge [-c]Z$) by direct exploration of the graph. The resulting tableau is, in the way of such things, quite large: it is shown in Figure 4.11.

This tableau is unsuccessful, owing to the terminal marked $*$ and its companion (also marked $*$): $\square_*$ is not well-founded. So the coverability graph fails to have the desired property. However, this tableau can be interpreted as a tableau for the net, by interpreting coverability graph nodes containing ω as sets of markings in the obvious way (namely N is the set

$$\{ M \mid \forall s \in \mathbf{S} . N(s) < \omega \Rightarrow M(s) = N(s) \});$$

and in this case, we can see that it is successful, since each passage through the rule of which the terminal $*$ is the consequent reduces the value of place B by 1. So although the tableau is bigger than the original 'inspired' tableau, the well-foundedness proof is actually easier, which is a pleasant bonus.

It should be noted that it is not automatically true that a tableau for a coverability graph translates to a tableau for the underlying net, let alone a successful tableau: if N has an a-successor for some net and graph, it is not necessarily true that every M covered by N has an a-successor, so a $\langle K \rangle$ rule in the coverability graph tableau may not translate to a valid rule in the net. The following proposition is the best that can be said about coverability graphs:

Proposition 4.33. Let $(\mathcal{N}, M_0)$ be a net and let G be a coverability graph. If Φ is a formula in the purely conjunctive fragment of the mu-calculus (that is, using only the connectives $\wedge$, $[K]$ and νZ.) such that any atomic propositions in Φ hold for N in G iff they hold for every marking in $\mathcal{N}$ represented by N, then a successful tableau for Φ in G can be translated to a successful tableau for Φ in $\mathcal{N}$, and hence if N satisfies Φ in G, any marking covered by N also satisfies Φ in $\mathcal{N}$.

Proof. The translation, almost trivial, is by induction on the structure of the tableau. The base case is dealt with by the condition on atomic

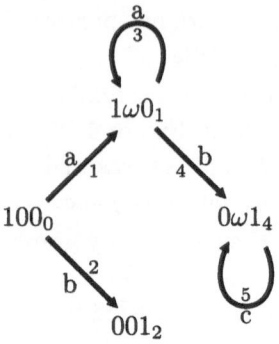

Figure 4.10

$$\frac{\{100\} \vdash \mu Y.\nu Z.[c]Y \wedge [-c]Z}{\{100\} \vdash U}$$

$$\frac{\{100\} \vdash \nu Z.[c]U \wedge [-c]Z}{\{100\} \vdash V}$$

$$\{100\} \vdash [c]U \wedge [-c]V$$

$$\frac{\{100\} \vdash [c]U}{\varnothing \vdash U} \qquad \frac{\{100\} \vdash [-c]V}{\{1\omega 0, 001\} \vdash V}$$

$$\{1\omega 0, 001\} \vdash [c]U \wedge [-c]V$$

$$\frac{\{1\omega 0, 001\} \vdash [c]U}{\varnothing \vdash U} \qquad \frac{\{1\omega 0, 001\} \vdash [-c]V}{\{1\omega 0, 0\omega 1\} \vdash V}$$

$$\{1\omega 0, 0\omega 1\} \vdash [c]U \wedge [-c]V$$

$$\frac{\{1\omega 0, 0\omega 1\} \vdash [c]U}{* \ \{0\omega 1\} \vdash U} \qquad \frac{\{1\omega 0, 0\omega 1\} \vdash [-c]V}{\{1\omega 0, 0\omega 1\} \vdash V}$$

$$\frac{\{0\omega 1\} \vdash \nu Z.[c]U \wedge [-c]Z}{\{0\omega 1\} \vdash V'}$$

$$\{0\omega 1\} \vdash [c]U \wedge [-c]V'$$

$$\frac{\{0\omega 1\} \vdash [c]U}{* \ \{0\omega 1\} \vdash U} \qquad \frac{\{0\omega 1\} \vdash [-c]V'}{\varnothing \vdash V'}$$

Figure 4.11

propositions. Assuming that as above a node of G denotes both itself and the set of markings it represents, the translation for the $\nu Z.$, $\wedge$, Un and Thin rule is the identity translation; this preserves the validity of the rule applications. This leaves only the $[K]$ rule: if in the G tableau we have

$$\frac{\{N_1, \ldots, N_n\} \vdash_\Delta [K]\Phi}{\{N'_1, \ldots, N'_{n'}\} \vdash_\Delta \Phi}$$

this is translated to

$$\frac{\{N_1, \ldots, N_n\} \vdash_\Delta [K]\Phi}{\mathbf{M} \vdash_\Delta \Phi}$$
$$\overline{\{N'_1, \ldots, N'_{n'}\} \vdash_\Delta \Phi}$$

where the $\mathbf{M}$ is the set of K-successors of $\{N_1, \ldots, N_n\}$ and the rule applied is Thin. To preserve validity of the rules we need to check that $\mathbf{M} \subseteq \{N'_1, \ldots, N'_{n'}\}$, so suppose the contrary, i.e. that there is a node N_i and a marking M represented by N_i such that $M \overset{t}{\longrightarrow} M'$ for some M' and M' is not represented by any N'_j. But this cannot happen, for either $M' \leq N$, in which case M' is also represented by N and by construction of G we have $N \overset{t}{\longrightarrow} N$, or else $N' = N - {}^\bullet t + t^\bullet$ appears in G and $N \overset{t}{\longrightarrow} N'$ in G.

Thus the translation preserves the validity of the rule applications, and since all U-sequents undergo the identity translation, the inclusion condition for ν-terminals remains true in the translated tableau. $\quad\square$

Finally, an example of how coverability graphs fail, without even using least fix-points. Consider the net of Figure 4.12 and the formula

$$[a]\nu Y.[-]Y \wedge [\overline{b}]\nu Z.\langle c\rangle \mathbf{tt} \wedge [c][c]Z$$

which says that after an a event then it is always true that once $\overline{b}$ fires there is no c sequence of even length, a property which is true of the initial marking shown.

The coverability graph of this net is shown in Figure 4.13, and as can be seen the formula above also holds of the node 100. Unfortunately, 100 also satisfies the formula with b substituted for $\overline{b}$, whereas the initial marking of the net does not. The reason for this is of course that the formula expresses a property which depends on the number of tokens on B being even, and so the coverability graph nodes are too crude a representation of sets of markings. One could invent generalizations of coverability graphs with slightly more expressive nodes, but the results of the next chapter

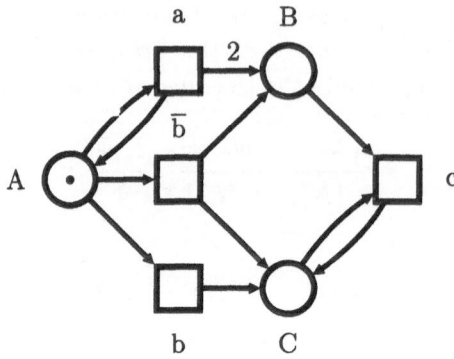

Figure 4.12

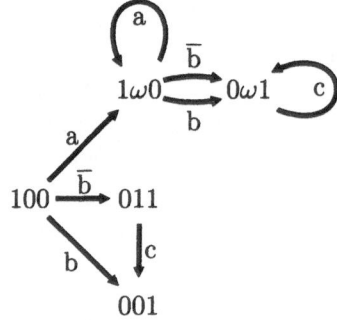

Figure 4.13

show that this could not work in general, so one must fall back on *ad hoc* exploration of the state space in the hope of noticing a pattern.

4.5 Some remarks on compositionality.

A fundamental problem with compositional approaches to model-checking is the product construction, be it CCS | or Petri net +. The properties of a product do not in general derive in any uniform way from the properties of the components. This point crops up time and again—it is forcefully made in [Win90] where very complicated reductions are used to get a partial solution.

Yet many systems have an intuitively obvious compositional structure, and moreover some properties of the system depend on properties of the components in a way that is intuitively comprehensible. This intuition should be formalizable. Therefore, we end this chapter by returning to the original reader–writer system and showing how the relation between invariants and our compositional notation for the system can be used to produce the crucial invariant for the safety property, without requiring the insight into the system that we previously assumed.

We first state the trivial but important

Proposition 4.34. The sum construction preserves and reflects invariants, that is, given nets $\mathcal{N}_1$ and $\mathcal{N}_2$, every pair ι_i of invariants of $\mathcal{N}_i$ gives an invariant $\begin{pmatrix} \iota_1 \\ \iota_2 \end{pmatrix}$ of $\mathcal{N}_1 + \mathcal{N}_2$, and every invariant of $\mathcal{N}_1 + \mathcal{N}_2$ projects to give invariants of $\mathcal{N}_1$ and $\mathcal{N}_2$. $\qquad\square$

Indeed, invariants are preserved by the Winskel-style synchronized product $\times_L$, which can be useful, but here we remain with the shared places approach. The problem is that the quotient construction does not in general preserve invariants—it reflects them:

Proposition 4.35. Given a net $\mathcal{N}$, and $\mathcal{N}' = \mathcal{N}/\beta$ for some β, if ι' is an invariant of $\mathcal{N}'$ then $\beta^{\mathrm{T}} \iota'$ is an invariant of $\mathcal{N}$.
Proof. Let N be the incidence matrix of $\mathcal{N}$, so βN is that of $\mathcal{N}'$. Then $\iota'^{\mathrm{T}}(\beta N) = 0 \Leftrightarrow (\beta^{\mathrm{T}} \iota')^{\mathrm{T}} N = 0$. $\qquad\square$

However, in the simple case of merging places, we can use this reflection to go the other way. Suppose we are merging places $s_1, \ldots, s_k$ into a single place s'. Then any invariant of $\mathcal{N}'$ is reflected in an invariant of $\mathcal{N}$ which is symmetrical in the s_i, and conversely if we can see an invariant of $\mathcal{N}$ which is symmetrical in the s_i, so has the form $a \sum_1^k s_i + \cdots$, we have that $as' + \cdots$ is an invariant of $\mathcal{N}'$. (So any invariant of $\mathcal{N}$ not involving the merged places is preserved in $\mathcal{N}'$.)

For example, take the reader–writer system as expressed earlier by

$$MRSW = \left(\sum_{i=1}^m Reader_i + \left(\sum_{j=1}^n Writer_j \right) / [Q_W = \{Q_{W_1}, \ldots, Q_{W_n}\}] \right)$$

$$/[Q_1 = \{Q_{R_1}, Q_W\}, \ldots, Q_m = \{Q_{R_m}, Q_W\}] / [Q = \{Q_1, \ldots, Q_m\}]$$

The invariants of the unquotiented system are very easy: all linear combinations of $(R_i + S_i)$, $(W_j + V_j)$, $(Q_{R_i} + R_i)$ and $(Q_{W_j} + W_j)$. Now the first quotient tells us to look for an invariant symmetrical in the Q_{W_j}, and the simplest non-trivial such is $\sum_j (Q_{W_j} + W_j)$, which gives us $Q_W + \sum_j W_j$ after quotienting. Then for each i we look for an invariant symmetrical in Q_{R_i} and Q_W: the simplest is $(Q_{R_i} + R_i) + (Q_W + \sum_j W_j)$, which after quotienting gives us $Q_i + R_i + \sum_j W_j$ for each i. Finally, the last quotient tells us to add these together, and after quotienting this gives $Q + \sum_i R_i + m \sum_j W_j$, which is the invariant we chose by 'insight' (together with $(R_i + S_i)$ etc., which carry through unchanged). Indeed, the reflection property tells us that these are the only invariants of the final system (up to linear combination).

Thus we have obtained a candidate invariant for our tableau purely from the composition of the very simple components. Of course, often this may not work—for one thing, linear invariants are not always adequate to describe sets of markings—but the principle is generally useful.

Chapter 5

The Complexity of Mu-Formulæ on Nets

To use the tableau system, one must have a means of representing sets of states. In the last chapter, we adopted linear inequalities as a presentation language, claiming that this is frequently sufficient. However, in the final example with coverability graphs, we saw a property which depended on a place having an even number of tokens, so showing that linear inequalities are certainly not sufficient even for small nets. The question naturally arises, what language suffices in all cases? That is, what is the expressive power required to represent all the sets of states that may be introduced in a tableau?

This question leads to the question, what is the expressive power of the mu-calculus in a given model? This second question does not immediately answer the first, since in general the appearance of a formula in a successful tableau does not require the ability to express the whole denotation of the formula—a subset may suffice, and that subset may be less complex than the whole set. However, a simple trick allows us, for Petri nets and similar models, to construct a formula such that the tableau must indeed contain the whole set, so we consider now this second question.

Note that this is not the same question as the general problem of the expressive power of the mu-calculus considered purely as a logic on transition systems, as in the work relating it to SnS, but rather with its expressive power within a certain class of models, which possess a rich structure enabling complex properties to be encoded in the mu-calculus.

The class of models we study is, of course, Petri nets. One reason is simply that having used Petri nets as examples, the answer to our question is interesting anyway, but even without that, Petri nets are an interesting choice: it turns out that although Petri nets are a strictly weaker model of computation than Turing machines, the modal mu-calculus can overcome this difference, and the expressive power is the same as that of the mu-calculus on non-deterministic Turing machines. It is also the case that the coding techniques required for these results are relatively straightforward and easy to visualize for Petri nets—we make much use of arithmetic, so it

is easier to work with Petri nets, which contain integers directly, than with
Turing machines with coding and tapes to worry about.

5.1 Beyond semi-linearity.

In section 4.4, we saw at the end a net and formula such that the denotation
of the formula depended on there being an even number of tokens on a
place. So we already know that linear inequalities are not enough even for
small nets. However, it is natural to wonder whether if we allow the use
of modular equations, we might achieve something. This is particularly
suggested by the use of semi-linearity in reachability analysis—it has been
shown that the reachability set of any net with fewer than six places is
semi-linear [HoP79]. (A set of vectors (markings) is semi-linear if every
element in it can be expressed as an offset, being a linear combination
from some fixed set of vectors, from one of a finite set of base points.) We
show now that semi-linearity does not suffice even for nets with only three
places. This is not of any great value in itself, but is quite amusing, and
both illustrates the main idea applied in the more important results and
serves as another example of the use of the tableau system.

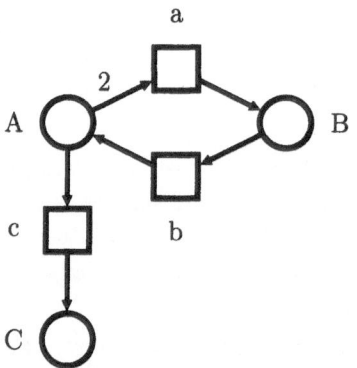

Figure 5.1

Consider the net in Figure 5.1. This is not very exciting *per se*, since
ultimately all it does is to move tokens from A to C, perhaps discarding
some of them. However, suppose we could control the order in which events
fire. In particular, suppose we could arrange that what happens is that a

fires for as long as possible, then c fires for as long as possible, then b for as long as possible, and then the cycle repeats. In that case, if we start with no tokens on B or C, each time round the cycle, A is replaced by $\lfloor A/2 \rfloor$, and A mod 2 is deposited on C. So the net terminates with C = 1 only if A was originally a power of 2. Of course, with Petri nets we cannot control events like this, but, and this is the key idea, we can use the mu-calculus to pick out only those paths which do cycle as above, and check for C = 1 at the end of them. Thus we get a mu-formula on the above net whose denotation requires exponentiation for its expression.

Proposition 5.1. Take the net of Figure 5.1 with any standard valuation. Let Φ be the formula

$$\mu X.(([-]\mathrm{ff} \wedge \mathrm{C} = 1) \vee [a]X) \vee ([a]\mathrm{ff}$$

$$\wedge \mu Y.[c]Y \vee ([c]\mathrm{ff} \wedge \mu Z.[b]Z \vee ([b]\mathrm{ff} \wedge X)))$$

Then $\|\Phi\| = \{(A, B, C) = (0, 0, 1)\} \cup \{C = 0 \wedge \exists n.\, A + 2B = 2^n\}$ (where the $\{\phi\}$ notation is as before).

Proof. Although calculating approximants is practicable in this case, still probably the easiest proof is to give a tableau to show that the given set satisfies Φ, and another to show that the complement of the set satisfies $\neg\Phi$.

Let ϕ be $(A, B, C) = (0, 0, 1)$, let ψ be $C = 0 \wedge \exists n.\, A + 2B = 2^n$ and let ψ' be $C = 0 \wedge \exists n.\, A + B = 2^n$ (the change to lower-case greek is to emphasize that these formulae are not being defined to be part of the mu-calculus). A tableau to show that $\{\phi \vee \psi\} \vDash \Phi$ is given in Figure 5.2. The application of the rules causes little difficulty: at 3 we thin to include the 001 marking since it appears at 5; the crucial step is at 6, where we thin and in so doing pass from looking at ψ to looking at ψ', which works because if $A + 2B = 2^n$ and $A = 0$ then $A + B = 2^{n-1}$; similarly, since $B = 0$, node 9 does satisfy the inclusion condition.

Well-foundedness is straightforward to show: for $\sqsubset_7$, take the measure B, which decreases by one each time through the rule above 8; and $\sqsubset_8$ is exactly $100 \sqsupset_8 001$. The non-trivial case is for $\sqsubset_1$, where we must take account of looping through the nested fix-points. First note that 001 does not go to anything, so we need only consider the non-ϕ part. A suitable measure is $(A + 2B, A + B)$: if we take the path from 1 to 2, the first component is constant and the second decreases by one, and if we take the extended path from 1 to 9, then either we take at least one loop through 8, in which case the first component decreases, or we take the direct path

Part 1

$$\frac{\{\phi \vee \psi\} \vdash \mu X.(([-]\text{ff} \wedge C = 1) \vee [a]X) \vee ([a]\text{ff} \wedge \mu Y....)}{{}_1\{\phi \vee \psi\} \vdash U}$$

$$\frac{\{\phi \vee \psi\} \vdash (([-]\text{ff} \wedge C = 1) \vee [a]U) \vee ([a]\text{ff} \wedge \mu Y....)}{\{\phi \vee (\psi \wedge A \geq 2)\} \vdash (([-]\text{ff} \wedge C = 1) \vee [a]U)} \qquad \text{Part 2}$$

$$\frac{\{\phi\} \vdash [-]\text{ff} \wedge C = 1 \qquad\qquad \{\psi \wedge A \geq 2\} \vdash [a]U}{\dfrac{\{\phi\} \vdash [-]\text{ff} \quad \{\phi\} \vdash C = 1 \quad {}_2\{\psi \wedge B \geq 1\} \vdash U}{\varnothing \vdash \text{ff}}}$$

Part 2

$$\frac{\{\psi \wedge A < 2\} \vdash [a]\text{ff} \wedge \mu Y....}{\dfrac{\{\psi \wedge A < 2\} \vdash [a]\text{ff} \quad \{\psi \wedge A < 2\} \vdash \mu Y.[c]Y \vee ([c]\text{ff} \wedge \mu Z....)}{\varnothing \vdash \text{ff} \qquad\qquad {}_3\{\psi \wedge A < 2\} \vdash V}}$$

$$\frac{{}_4\{\phi \vee (\psi \wedge A < 2)\} \vdash V}{\{\phi \vee (\psi \wedge A < 2)\} \vdash [c]V \vee ([c]\text{ff} \wedge \mu Z....)}$$

$$\frac{\{\psi \wedge A = 1\} \vdash [c]V \qquad \{\phi \vee (\psi \wedge A = 0)\} \vdash [c]\text{ff} \wedge \mu Z....}{{}_5\{\phi\} \vdash V \qquad\qquad \dfrac{\{\phi \vee (\psi \wedge A = 0)\} \vdash [c]\text{ff} \qquad \text{Part 3}}{\varnothing \vdash \text{ff}}}$$

Part 3

$$\frac{\{\phi \vee (\psi \wedge A = 0)\} \vdash \mu Z.[b]Z \vee ([b]\text{ff} \wedge U)}{{}_6\{\phi \vee (\psi \wedge A = 0)\} \vdash W}$$

$$\frac{{}_7\{\phi \vee \psi'\} \vdash W}{\{\phi \vee \psi'\} \vdash [b]W \vee ([b]\text{ff} \wedge U)}$$

$$\frac{\{\psi' \wedge B \geq 1\} \vdash [b]W \qquad\qquad \{\phi \vee (\psi' \wedge B = 0)\} \vdash [b]\text{ff} \wedge U}{{}_8\{\psi' \wedge A \geq 1\} \vdash W \qquad \dfrac{\{\phi \vee (\psi' \wedge B = 0)\} \qquad {}_9\{\phi \vee (\psi' \wedge B = 0)\}}{\vdash [b]\text{ff} \qquad\qquad \vdash U}}$$

$$\varnothing \vdash \text{ff}$$

Figure 5.2

from 1 to 9, in which case we have B = 0 (from the set at 9) and A < 2 (from 3), and therefore by ψ we get A = 1, which contradicts the fact that A = 0 (from 6).

The tableau to show that $\{\neg(\phi \vee \psi)\} \vDash \neg\Phi$ is left to the reader: it can be built from this tableau by dualizing—this tableau is canonical, so negate all the sets at the companions to get the greatest fix-point sets, and then make the rest consistent with this, using Thin as necessary. $\square$

5.2 Undecidability of the model-checking problem.

Now let us turn to the general case. The first major result is simply that the model-checking problem on Petri nets is in general undecidable, even for some formulae with only one fix-point operator. This result is shown by a conceptually simple construction: just as we used the mu-calculus to select those paths in the previous net which simulate repeated halving, so we can build a net representing a register machine and select only the paths corresponding to an execution of the machine. Thus, we can encode the halting problem.

Definition 5.2. Let $\mathcal{R}$ be a register machine, that is, a tuple

$$(\{q_0, \ldots, q_n\}, \{R_1, \ldots, R_m\}, \{\delta_0, \ldots, \delta_{n-1}\})$$

where R_i are the registers, q_i are the states with q_0 being the initial state and q_n the unique halting state, and δ_i is the transition rule for state q_i: δ_i is either (i) '$R_j := R_j + 1$; goto q_k' for some j, k, or (ii) 'if $R_j = 0$ then goto q_k else ($R_j := R_j - 1$; goto $q_{k'}$)' for some j, k, k'.

We define a net $\mathcal{N} = \text{net}(\mathcal{R})$ thus: the places of $\mathcal{N}$ are $q_0, \ldots, q_n, R_1, \ldots,$ R_m. The transitions and flow relation are determined by the δ_i: if δ_i is of form (i), then there is a transition δ_i^+ with ${}^\bullet\delta_i^+ = \{q_i\}$ and $\delta_i^{+\bullet} = \{q_k, R_j\}$; and if δ_i is of form (ii), then there are transitions δ_i^0 and δ_i^- such that ${}^\bullet\delta_i^0 = \{q_i\}$, ${}^\bullet\delta_i^- = \{q_i, R_j\}$, $\delta_i^{0\bullet} = \{q_k\}$ and $\delta_i^{-\bullet} = \{q_{k'}\}$. This is shown graphically in Figure 5.3. ◁

Definition 5.3. Let $\mathcal{N} = \text{net}(\mathcal{R})$ be as above. Let Δ^+ (resp. Δ^0, Δ^-) be the set of δ^+ (resp. δ^0, δ^-) transitions, let $\Delta^\pm = \Delta^+ \cup \Delta^-$, and $\Delta = \Delta^\pm \cup \Delta^0$.

Define the mu-formula $\text{halt}(\mathcal{N})$ to be

$$\mu Z.\left(\sum_{i=0}^{n} q_i = 1\right) \wedge [\Delta^\pm]Z \wedge ([\Delta^\pm]\text{ff} \Rightarrow [\Delta^0]Z)$$

◁

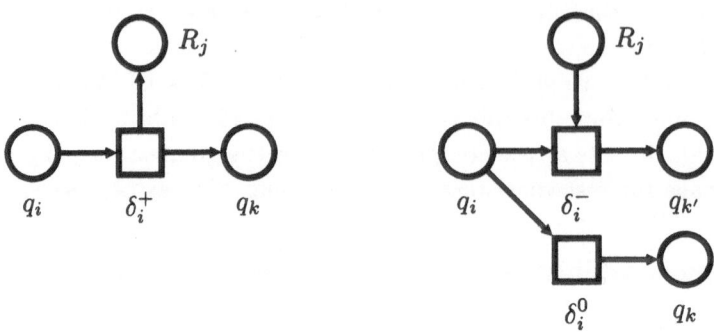

Figure 5.3

Theorem 5.4. For $\mathcal{N} = \text{net}(\mathcal{R})$ and a marking M of $\mathcal{N}$, $M \vDash \text{halt}(\mathcal{N})$ iff M corresponds to a configuration of $\mathcal{R}$ (i.e. $\sum_{i=0}^{n} q_i = 1$) and $\mathcal{R}$ halts when started in that configuration.

Proof. Consider the canonical tableau for $M \vdash \text{halt}(\mathcal{N})$, shown in Figure 5.4 (with the state sets left unspecified). Firstly, note that if M is a valid configuration, so are all markings reachable from M, since $\sum_{i=0}^{n} q_i$ is an invariant. Now suppose that $M \sqsupset_1 M'$. If $M@1 \rightarrowtail M'@2$, this happens by means of either a δ^+ or a δ^- transition, and in either case this corresponds to a valid transition in the machine. If on the other hand $M@1 \rightarrowtail M'@4$, then M corresponds to a configuration in state q_i, some i, and the associated δ_i^0 transition fires. This is only a valid transition in the machine if $R_j = 0$, where $^\bullet\delta_i^- = \{R_j\}$. However, since the tableau is canonical, the disjunction at 3 ensures that this is so, since otherwise δ_i^- could fire and M would go into the left disjunct. Also, if M corresponds to a valid machine configuration in which the machine can execute an instruction to go to a state corresponding to M', then $M \sqsupset_1 M'$.

Therefore, the well-foundedness of $\sqsubset_1$ implies that a machine execution sequence starting at M terminates. Conversely, if $\mathcal{R}$ terminates when started at M, a successful tableau for $M \vdash \text{halt}(\Phi)$ can be built just by weakening to the reachability set of $(\mathcal{N}, M)$ at the constant introduction, and applying the tableau rules, ensuring that the left disjunct at 3 contains all those markings satisfying $\langle \Delta^{\pm} \rangle \text{tt}$; then the execution relation corresponds to $\sqsupset_1$ as before.   □

Corollary 5.5. There is a net $\mathcal{N}$ for which $\text{halt}(\mathcal{N})$ is undecidable.   □

$$\frac{M \vdash \mu Z.\left(\sum_{i=0}^{n} q_i = 1\right) \wedge \left([\Delta^{\pm}]Z \wedge \left(\langle\Delta^{\pm}\rangle \mathrm{tt} \vee [\Delta^0]Z\right)\right)}{\dfrac{M \vdash U}{_1- \vdash U}}$$

$$\frac{- \vdash \left(\sum_{i=0}^{n} q_i = 1\right) \wedge \left([\Delta^{\pm}]U \wedge \left(\langle\Delta^{\pm}\rangle \mathrm{tt} \vee [\Delta^0]U\right)\right)}{- \vdash \left(\sum_{i=0}^{n} q_i = 1\right) \qquad \dfrac{- \vdash [\Delta^{\pm}]U \wedge \left(\langle\Delta^{\pm}\rangle \mathrm{tt} \vee [\Delta^0]U\right)}{\dfrac{- \vdash [\Delta^{\pm}]U}{_2- \vdash U} \qquad \dfrac{_3- \vdash \langle\Delta^{\pm}\rangle \mathrm{tt} \vee [\Delta^0]U}{\dfrac{- \vdash \langle\Delta^{\pm}\rangle \mathrm{tt}}{- \vdash \mathrm{tt}} \qquad \dfrac{- \vdash [\Delta^0]U}{_4- \vdash U}}}}$$

Figure 5.4

It should come as no surprise that the model-checking problem for nets is not in general decidable: there are many other ways of showing this. For example, the problem of whether the reachability set of one net is included in that of another has been known for some time to be undecidable (Rabin, given in [Hac76]), and by a similar coding and path selection trick this problem can be turned into a model checking problem on one net. Similar things could be done with other results from the large literature on the low-level complexity of Petri nets. However, the theorem we have just proved, as well as providing a very simple undecidable model-checking problem, also serves as a springboard from which we can jump to the high-level complexity required for general mu-calculus model-checking.

5.3 Ascending the arithmetical hierarchy.

We now know that we can express some undecidable sets, so our presentation language for marking sets must in general be strong enough to do this. Now the question is just how bad can things get? This section continues the exploration by pushing up the lower bounds on the complexity of mu-formulae.

We start by recalling some basic definitions and facts from recursion theory and effective descriptive set theory. As usual in this area, relations are sets, and functions are sets by considering their graphs as relations.

Definition 5.6. A relation on the natural numbers is Π_n or Σ_n if it is definable in the first-order language of arithmetic by a formula in prenex normal form with n alternating blocks of quantifiers, the outermost block being $\forall$ or $\exists$ respectively (so Σ_n is the complement of Π_n). It is Δ_n if it is both Π_n and Σ_n.   ◁

Fact 5.7. (i) The hierarchy Σ_n is a strict hierarchy, called the arithmetical hierarchy.

(ii) (Matijacevič' theorem) The set of Σ_1 sets is exactly the set of recursively enumerable sets (and therefore the Δ_1 sets are the recursive sets). If we define Σ_n over the language with constants for all recursive relations (rather than that of arithmetic), the hierarchy is unchanged above Σ_0.   □

Definition 5.8. Similarly, a relation is Σ_n^1 (Π_n^1) if it is definable in the second-order language of arithmetic by a formula in prenex normal form with n alternating blocks of second-order quantifiers, the outermost being $\exists$ ($\forall$), followed by a first-order formula. The hierarchy Σ_n^1 is called the analytical hierarchy.   ◁

We now extend the halting formula to specify a condition that should hold on termination. Remember that we are assuming standard valuations, so we have linear inequalities as atomic propositions.

Definition 5.9. For a net $\mathcal{N} = \mathrm{net}(\mathcal{R})$ as above, and a mu-formula Φ, the formula $\mathrm{halt}(\mathcal{N}, \Phi)$ is defined to be

$$\mu Z.\Big(\sum_{i=0}^{n} q_i = 1\Big) \wedge (([\Delta]\mathrm{ff} \wedge \Phi) \vee ([\Delta^{\pm}]Z \wedge ([\Delta^{\pm}]\mathrm{ff} \Rightarrow [\Delta^0]Z)))$$

(where Z is not free in Φ).   ◁

Just as above, $\mathrm{halt}(\mathcal{N}, \Phi)$ is satisfied by exactly those markings corresponding to configurations of $\mathcal{R}$ in which it halts and on termination satisfies Φ. This allows us to express recursive relations with nets and mu-calculus.

Proposition 5.10. Let S be a recursive (unary) relation on N. Then there is a net $\mathcal{N}_S$ and a μ_1 mu-formula Φ_S, and places q_0 and R_1 of the net such that $\|\Phi_S\|$ is exactly those markings which have all places zero save that $q_0 = 1$ and $R_1 = n$ for some n such that $n \in S$.

Proof. Let $\mathcal{R}_S$ be a register machine which computes the characteristic function of S, with the input in R_1 and the answer being left in R_1. Then $\mathcal{N}_S$ is $\mathrm{net}(\mathcal{R}_S)$, and Φ_S is $\Phi_{\mathrm{init}} \wedge \mathrm{halt}(\mathcal{N}_S, R_1 = 1)$, where $\Phi_{\mathrm{init}} = (q_0 = 1) \wedge \bigwedge_{i=2}^{m}(R_i = 0)$; this formula is μ_1.   □

Note that second clause of Φ_{init} is not really necessary, since one can initialize registers to any desired value by extending the machine slightly. Note also that by simple coding we can deal with n-ary relations, or alternatively use n input registers.

Since the recursive relations are closed under negation, we can also get a version of Proposition 5.10 with ν_1 formulae:

Proposition 5.11. Proposition 5.10 remains true if instead we require a net $\mathcal{N}'_S$ and a ν_1 formula Φ'_S.
Proof. Let $\mathcal{N}'_S$ be $\mathcal{N}_{\neg S}$ and let Φ'_S be $\Phi_{\text{init}} \wedge \neg\text{halt}(\mathcal{N}'_S, R_1 = 1)$. $\qquad\square$

Now we combine this construction with a token generator to ascend the arithmetical hierarchy.

Theorem 5.12. If S is a (unary, wlog) relation defined by an arithmetical formula ϕ, then there is a net $\mathcal{N}_S$ and a mu-formula Φ_S such that $\|\Phi_S\|$ is the set of markings such that all places are empty save that $q_0 = 1$ and $R_1 = n$ for some n such that $n \in S$, and furthermore, if S is Σ_l (resp. Π_l) then Φ_S is μ_l (resp. ν_l).
Proof. Let ϕ have the form $Q_1 x_1 \ldots Q_l x_l . \psi$, where each Q_j is either $\forall$ or $\exists$, and ψ is recursive. Construct a net $\mathcal{N}_\psi$ and mu-formula Φ_ψ as above, with the input places being R_0 (corresponding to the free variable in ϕ) and $R_1, \ldots, R_l$ (corresponding to the variables $x_1, \ldots, x_n$); choose the μ_1 form or the ν_1 form according as Q_l is $\exists$ or $\forall$.

Now construct $\mathcal{N}_S$ thus: the places of $\mathcal{N}_S$ are the places of $\mathcal{N}_\psi$ together with a place A_j for each quantified variable x_j. The transitions and flow relation are those of $\mathcal{N}_\psi$ extended by, for each variable x_j, transitions a_j and b_j such that

$$^\bullet a_j = \{A_j\} \quad a_j{}^\bullet = \{A_j, R_j\}$$

$$^\bullet b_j = \{A_j\} \quad b_j{}^\bullet = \{A_{j+1}\} \quad \text{for} \quad j < l$$

$$^\bullet b_l = \{A_l\} \quad b_l{}^\bullet = \{q_0\}$$

This is illustrated graphically in Figure 5.5.

Define

$$\text{quant}_j(\Psi) = \begin{cases} \mu Z_j . \langle a_j \rangle Z_j \vee \langle b_j \rangle \Psi & \text{if } Q_j \text{ is } \exists \\ \nu Z_j . [a_j] Z_j \wedge [b_j] \Psi & \text{if } Q_j \text{ is } \forall \end{cases}$$

and let $\Phi_{\text{init}} = (A_1 = 1$ and all other places apart from R_0 empty). Then the formula Φ_S is

$$\Phi_{\text{init}} \wedge \text{quant}_1(\text{quant}_2(\ldots \text{quant}_l(\Phi_\psi) \ldots)).$$

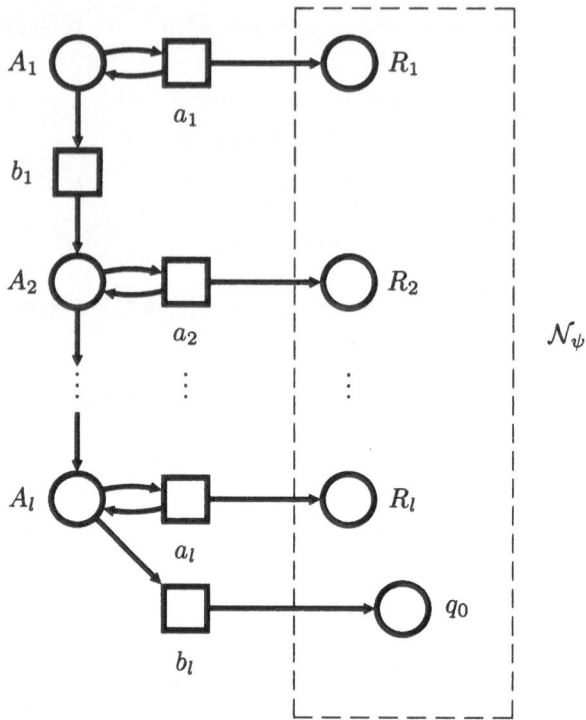

Figure 5.5

Suppose $M_j \vDash \text{quant}_j(\ldots)$, with $M_j(A_j) = 1$ and $M_j(A_k) = 0$ for $k \neq j$. If Q_j is $\exists$, then by definition of quant_j there is a marking M_{j+1} such that $M_j \xrightarrow{a_j}^* M_{j+1}$ and $M_{j+1} \vDash \langle b_j \rangle \text{quant}_{j_1}(\ldots)$, which is to say, $\exists r_j$. $M' \vDash \text{quant}_{j_1}(\ldots)$, where M' is the same as M_j except that $M'(A_j) = 0$, $M'(A_{j+1}) = 1$ and $M'(R_j) = M_j(R_j) + r_j$. Similarly, if Q_j is $\forall$, then $\forall r_j$. $M' \vDash \text{quant}_{j+1}(\ldots)$.

So, if $M \vDash \Phi_S$, we have that

$$Q_1 r_1 \ldots Q_l r_l . M' \vDash \Phi_\psi$$

where here $M'(R_j) = r_j$, $M'(q_0) = 1)$, $M'(R_0) = M(R_0)$, and all other places are empty. Now by Proposition 5.10, $M' \vDash \Phi_\psi$ iff $\psi(r_0, r_1, \ldots, r_l)$.

The converse, that if $n \in S$ the corresponding marking satisfies Φ_S is similarly straightforward, either directly or by dualizing.

Finally, Φ_S is indeed in μ_l or ν_l as required. $\square$

Now we have established that an arithmetical set is expressible in some net by a mu-formula at a corresponding level in the hierarchy, we naturally wish to prove the converse. This is done by a routine arithmetization of the modal connectives, followed by coding up the idea of approximants. First, a lemma that allows the coding:

Lemma 5.13. For any net constructed as above, all fix-point subformulae of a μ_n or ν_n mu-formula have closure ordinal ω.

Proof. Since the transition system determined by a finite Petri net is finite-branching, and any fix-point subformulae of a μ_n or ν_n formula is a fix-subsentence, this lemma is just Proposition 2.18. $\square$

Theorem 5.14. For any net $\mathcal{N}$, the denotation of a μ_n (resp. ν_n) mu-formula is Σ_n (resp. Π_n).

Proof. The quantification in the modal connectives is not a problem, since all our nets are finite and we have an interleaving semantics. (Allowing sets of transitions to fire simultaneously is unproblematic; the difficulty comes with multisets, and then only if the transition with no preconditions is allowed.) We arithmetize the notion of belonging to a finite unfolding thus:

Let Φ be a fix-point-free formula with one free variable Z. Let Z also be a unary predicate symbol in arithmetic. We define a formula ϕ recursive in Z by induction on the structure of Φ.

- if Φ is atomic or Z, $\phi(M)$ is $\hat{\Phi}(M)$ or $Z(M)$, where $\hat{\Phi}$ is a constant interpreting Φ;
- the boolean combinators carry over unchanged;
- if $\Phi = [K]\Psi$, then $\phi(M) \stackrel{\text{def}}{=} \bigwedge_{a \in K}(\hat{a}(M) \Rightarrow \psi(\tilde{a}(M)))$, where ψ is the translation of Ψ, $\hat{a}$ is the (recursive) predicate interpreting a is enabled at M, and $\tilde{a}$ is the (recursive) function such $\tilde{a}(M)$ is the unique M' such that $M \stackrel{a}{\longrightarrow} M'$
- if $\Phi = \langle K \rangle \Psi$, then $\phi(M) \stackrel{\text{def}}{=} \bigvee_{a \in K}(\hat{a}(M) \wedge \psi(\tilde{a}(M)))$.

Thus $\phi(M)$ is a direct coding of $M \vDash \Phi$, if the arithmetic Z interprets the mu-calculus Z. In particular, if we now define

$$f(M, 0) = 0$$

$$f(M, n+1) = \phi(M)[f(-, n)/Z(-)]$$

we have a recursive function $f(M, n)$ which is the characteristic function of $M \vDash \mu^n Z.\Phi$. Therefore, by the previous lemma we have that $M \vDash \mu Z.\Phi$ iff $\exists n \, . \, f(M, n)$. Dually for $\nu Z.$, and for fix-point subformulae, we simply use the translation of the subformula. Putting the resulting translation in prenex normal form gives the result. (This last stage corresponds exactly to the definition of $\hat{\mu}_n$.) $\square$

So we now know exactly the expressive power of unnested mu-formulae on nets. At the beginning of the chapter, we mentioned the distinction between expressing the denotation of a formula and being able to express a tableau, and said that it did not matter; we should justify this. It is not quite true that it makes no difference: for example, while a μ_1 formula may have a Σ_1 denotation, any finite subset of that denotation has a recursively presentable tableau, since it is contained in some finite unfolding. However, at the cost of another couple of fix-point levels, we can force the entire denotation to appear in the tableau: take the construction for some arithmetical formula ϕ and just add another token generator, with a transition c say, which adds tokens to the input register R_0 of our net, and is disabled as soon the generator for the first quantified variable starts, and consider 'on all c-runs, always eventually Φ_ϕ'; a tableau for this formula must necessarily include the whole denotation of Φ_ϕ.

So to sum up the results of this section,

Theorem 5.15. (i) For any net $\mathcal{N}$ (and sensible valuation, i.e. one that assigns recursive properties to variable symbols), if a mu-formula Φ is μ_n (resp. ν_n), then $\|\Phi\|$ is Σ_n (resp. Π_n).

(ii) For any arithmetical set S which is Σ_n (resp. Π_n), there exists a net $\mathcal{N}$ and a mu-formula Φ such that $\|\Phi\| = S$ and Φ is μ_n (resp. ν_n). $\square$

5.4 Beyond the arithmetical hierarchy.

The last section dealt only with unnested fix-point formulae. It is natural to wonder whether, as the μ_n hierarchy corresponds to the arithmetical hierarchy, the μ_n hierarchy might correspond to the analytical hierarchy. It is obvious that the denotation of a μ_n formula is Σ_n^1—just code up the formula in the way used above, and write out the definition of minimal and maximal fix-points. The problem of whether this expressive power is attainable, as well as being an upper bound, is something about which it is very difficult to have any intuition—it might seem plausible, or it might seem unlikely that the fix-point operators could give the full power of second-order quantification. Unfortunately, so far we have not been able to give any satisfactory answer. This is not really surprising, since the problem is related to similar problems in mathematics, namely the field of inductive and co-inductive definitions, in which the questions are also still open. However, we can say that alternating formulae require more than arithmetical power. The proofs of this again proceed by coding tricks in the style of the previous section; the formal details of the coding are much as before, so to avoid excessive repetition of tedious notation we shall just give a higher-level description; any reader who wishes, will have no difficulty in writing out details.

We state the theorem, and then go on to discuss informally two proofs, each connecting our problem to an established area of mathematics.

Theorem 5.16. The complexity of sets expressible by the mu-calculus on nets is at least $\Sigma_1^1 \cup \Pi_1^1$. □

5.4.1 Proof by inductive definitions.

In definability theory, a relation S is said to be positive elementary inductive if it can be defined as the least fix-point of an arithmetical sentence with one free predicate variable occurring positively, and this can be expressed in terms of approximants exactly as for the mu-calculus least fix-point. So it is an obvious ploy to try to relate this to a least fix-point formula in the mu-calculus, and indeed, this can be done. (See [Acz77] for an introduction to inductive definitions.)

The easiest way to think about the construction is to imagine adding oracles to our nets representing machines. Suppose we have some arithmetical formula ϕ with Z being a free predicate symbol occurring positively. By the techniques of the previous section, we can build a net which 'computes' ϕ, provided it can be told the answer to the membership question for Z. So we arrange that when the net/machine wishes to know whether $n \in Z$, it puts n in a designated register, and enters a special state q_Z. At this point, a magic oracle decides whether $n \in Z$ and changes the state to, say, $q_{\in Z}$ or $q_{\notin Z}$ accordingly. We model this just by adding two transitions to the net, say, z_1 and z_0, which transfer the state token; the oracle's job is then to choose which transition fires according to whether $n \in Z$. Now, if we could express Z by some mu-formula Φ_Z, we could select the valid computations of this machine with an oracle just by extending the innermost halt formula with a disjunct

$$(q_Z = 1 \wedge ((\Phi_Z \wedge [z_1]Z) \vee (\neg\Phi_Z \wedge [z_2]Z))).$$

The trick is simply to observe that since we want a machine to calculate the least fix-point of ϕ, the machine is its own oracle. So, what we is have (another!) special transition t from the state q_Z, which transfers control to a state which reinitializes the machine and restarts it on the value n. So to get Φ_Z, all we do is wrap Φ_ϕ up in a $\mu Z.$, and make the above addition to the inner halt formula, replacing Φ_Z by $[t]Z$. One detail was skipped: we cannot use $\neg[t]Z$ to represent $\neg\Phi_Z$, since then Z would occur negatively. However, since ϕ is positive, and therefore monotonic, in Z, we can safely omit this term altogether—if n is calculated to be in Z based on n' wrongly being calculated not to be in Z, then n really is in Z, by monotonicity.

At this point, we apply Kleene's theorem, that the positive elementary inductive sets (over arithmetic) are exactly the Π_1^1 sets, to get the result.

One might wonder whether this technique can be continued to get yet more complex sets. Unfortunately, as far as I can determine from the mathematical literature, the complexity of induction over non-arithmetical sets is not known in terms of the analytical hierarchy, and indeed very little appears to be known about combining induction and co-induction (corresponding to greatest fix-points). Thus the continued investigation of the expressive power of the mu-calculus in nets connects intimately with a very rich and beautiful area of mathematics, and for that reason alone is submitted as a worthy topic of future research, even were it of no interest in itself.

5.4.2 Proof by partially-ordered quantification.

The second approach again relies on results from mathematical logic, and again, the problems are still open.

When a first order formula is expressed in prenex normal form, an inner variable depends on all the variables before it—when playing the quantifier game, values of variables are chosen from the outside in. A possible extension of first order logic is to consider quantifiers with a partial, rather than linear, dependency ordering. A simple example is the *Henkin quantifier*, usually written $\genfrac{}{}{0pt}{}{\forall x.\,\exists y.}{\forall v.\,\exists w.}$, in which the choice of y depends only on x, and the choice of w depends only on v. These were first studied by Ehrenfeucht and Henkin, and it was shown that the addition of the Henkin quantifier allows the expression of the cardinality quantifier 'there exist at least $\aleph_0$ x', and therefore that it strictly extends the expressive power of first-order logic.

Now, just as we designed nets for normal first-order formulae, so we can design nets for partially ordered quantifiers, at least of a certain well-behaved form (namely, those that can be written in rows like the Henkin quantifier, i.e. the dependency partial order is a union of disjoint linear orders). We demonstrate the construction of a net and mu-formula for a Henkin quantified formula; the general construction is similar.

Consider the formula $\overset{\forall x \,.\, \exists y \,.}{\forall v \,.\, \exists w \,.}\, \phi$, where ϕ is arithmetical. Let $\mathcal{N}_\phi$ be the net constructed for ϕ as in the previous section, with input places R_x etc. for the quantified variables. Construct the net $\mathcal{N}$ shown in Figure 5.6.

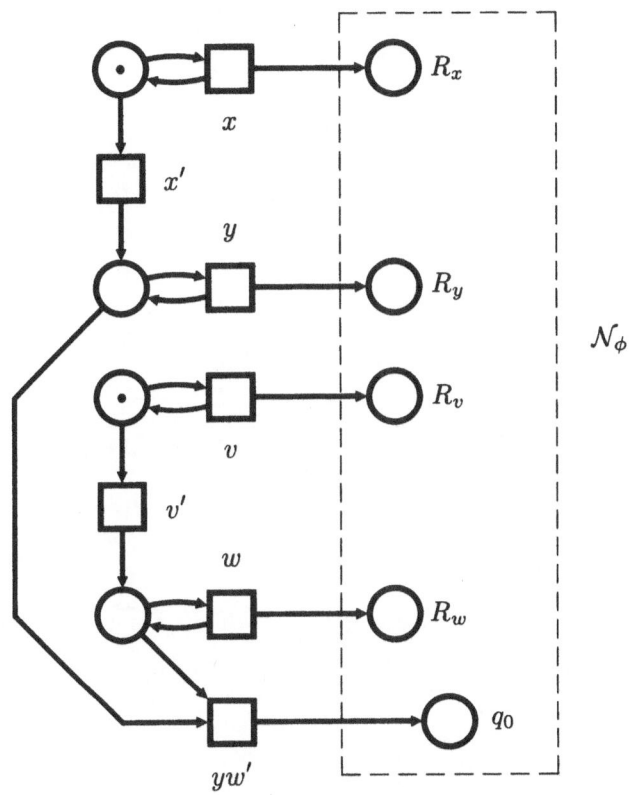

Figure 5.6

The mu-formula that describes the markings satisfying $\overset{\forall x \,.\, \exists y \,.}{\forall v \,.\, \exists w \,.}\, \phi$ is then

$$\nu X.\mu Y.[x,v]X \land (\langle yw' \rangle \Phi_\phi \lor \langle x', v', y, w \rangle Y)$$

plus the usual initialization formula.

So, we can code partially quantified formulae into mu-formulae. This gives us the result, since by a theorem of Walkoe [Wal70] and Enderton (independently), any Σ_1^1 formula in the language of arithmetic is equivalent to a partially quantified first order formula (and indeed, this is still true if *only* the Henkin quantifier is allowed as a partial quantifier).

Unfortunately, it is again an open problem how great is the expressive power of partially-ordered quantification, so this technique does not give us any further information. However, it is yet another avenue for further research.

Chapter 6

Conclusions and Further Work

The verification of infinite systems, or even large finite systems, is a task of great importance, and becomes ever more so as systems become more parallel and more distributed. In this monograph I have presented a general framework for proving a wide range of properties of systems. Because I have used a logic with very few primitives, the system is conceptually rather simple; but since the logic is very expressive, the system is powerful. This power has the inevitable consequence, as we saw in the last chapter, that in general the system is not effective. Indeed, although the system is complete in a model-theoretic sense, in that for every true property there exists within the set-theoretic universe a successful tableau for that property, the ability to code arithmetic shows that there is no complete formal proof system for the success of tableaux upon any reasonably powerful model of computation.

These results are inherent in the nature of infinite-state model-checking, and emphasize how different it is from the finite case. There are many questions still to be asked and answered for infinite-state model-checking: let us now consider a few of these.

6.1 Incorporating reasoning.

Owing to the non-effective nature of the tableau system, I see it as something to be used by humans, albeit with (I hope) considerable computer assistance, rather than by machines. Moreover, because the system allows, and frequently needs, the use of reasoning specific to some class of models or even some particular model, there is much scope for work on the use of, say, mechanical theorem-provers with the tableau system. For example, in chapter 4 we made much use of the invariant calculus and of linear inequalities. We should therefore like an implementation to be able to manipulate such expressions, and so partially to automate the construction of tableaux. The general problem is to what extent it is (a) possible and (b) desirable to formalize the use of various techniques within the tableau system; this is interesting and open-ended, and I look forward to experimenting in this area.

6.2 Decidability of model-checking.

Although I believe in the use of human provers, it is of course interesting and useful to know under what circumstances there is an effective method of verifying a system. The results of chapter 5 show that in general there is no hope of effective proofs. However, there may be certain classes of models or properties for which model-checking is decidable.

For example, we already know that a rather uninteresting class of properties, those expressible in the conjunctive fragment of the mu-calculus, are decidable on Petri nets (Proposition 4.33); can this be enlarged? Or, what is the effect of placing various limitations on atomic propositions?

More promisingly, are there particular classes of processes or nets for which all formulae can be effectively checked? A process class that is a prime candidate for this is the class of context-free processes: Hüttel and Stirling [HüS91] have used a tableau proof method to show that bisimilarity of such processes is decidable, in a rather more intuitive way than previous proofs of this. So the question of model-checking context-free processes using tableau techniques is an obvious candidate for investigation. Petri nets are also a rich field for this sort of investigation: we mentioned in chapter 4 a few properties of nets expressible in the modal mu-calculus, and there are plenty more. Indeed, if we add backwards modalities $\Phi[K]$ (defined in the obvious way by $s \vDash \Phi[K]$ iff $\forall s' . s' \xrightarrow{K} s \Rightarrow s' \vDash \Phi$), we can express the reachability set of $(\mathcal{N}, M_0)$ as $\mu Z.M_0 \vee Z\langle - \rangle$. Questions of this kind have very much the feel of local model-checking: they are not about the logic as such (one could as well add backwards transition relations to the model as add backwards modalities to the logic), but about the interaction of the logic and the model.

6.3 Proving success.

Another issue related to this is how to prove well-foundedness. In chapter 5, we were concerned simply with expressing tableaux, but we did not consider how to write down proofs of well-foundedness. The example of home states, or the reachability set, shows that quite simple tableaux may have far from simple well-foundedness proofs. So we may ask, which tableaux can we prove successful by using a given formal proof system? Even if we allow ourselves a strong system such as Peano Arithmetic, it is not difficult (ignoring practical difficulties such as writing down a correct register-machine program!) to build a net of some eighty elements, which on selected execution sequences always terminates, but such that its termination cannot be proven in Peano Arithmetic.

This can be done by encoding the Goodstein function. This function (in a more transparent presentation than the original) is a function between pairs (α, n), where α is an ordinal less than ϵ_0 and n is an integer. First define an auxiliary function $d: \epsilon_0 \times \mathsf{N} \to \epsilon_0$ (d for 'decrement') as follows, where α is assumed to be written in canonical form (i.e. as a sum of decreasing powers of ω):

$$d(0, n) = 0$$

$$d(m, n) = m - 1$$

$$d(\alpha + \beta, n) = \alpha + d(\beta, n)$$

$$d(\omega^\alpha, n) = d(\omega^{d(\alpha, n)} \cdot n, n)$$

and then define the Goodstein function g by

$$g(0, n) = (0, n)$$

$$g(\alpha, n) = g(d(\alpha, n + 1), n + 1)$$

This function always terminates, and in the presentation here it is fairly easy to see why: as suggested by its name, the function d always strictly decreases the ordinal. It is also clear that using register machine coding tricks in the style of chapter 5, we can build a Petri net to simulate execution of this function, and express the termination property. However, the termination proof relies on induction up to ϵ_0, and it is known that this reliance is essential; and since ϵ_0-induction is not derivable in Peano Arithmetic, termination of our net cannot be proven in Peano Arithmetic. (Moreover, no brute force exploration will show that it does: a little calculation shows that $g(\omega^\omega, 2)$ terminates after roughly 7×10^{121210694} iterations! This looks even more impressive in the original formulation, which represents α as an integer by evaluating the ω-polynomial representation taking ω to have value n, so we would say $g(4, 2)$, and describes the function as "write an integer (corresponding to α) as a sum of powers of n (the exponents also being written in this form, and so on), replace all occurrences of n by $n+1$, subtract one from the result and repeat the whole procedure".)

This example points out the difficulty of restricting the complexity of mu-calculus properties: there is nothing obvious about the net to make its termination so hard to prove. It seems likely that quite severe restrictions will have to be put upon nets to constrain them to have nice properties for model-checking. For a final example, we present a six-element net and a μ_1 formula whose denotation is probably decidable, and in fact is generally believed to the whole marking set, but for which there is as yet no proof of either of these statements.

Consider the net

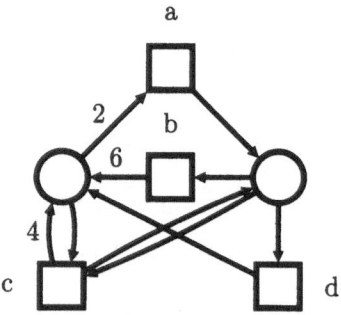

and the formula

$$\mu X.[-]\text{ff} \vee \langle a \rangle X \vee ([a]\text{ff} \wedge (\langle c \rangle(\mu Z.\langle b \rangle Z \vee ([b]\text{ff} \wedge X))$$

$$\vee \; ([c]\text{ff} \wedge \mu Z.\langle d \rangle Z \vee ([d]\text{ff} \wedge X))))$$

The reader who has followed chapter 5 will see that this net and formula encode the famous $3n+1$ problem: is the partial recursive function f given by

$$f(n) = \text{if } n \leq 1 \text{ then } n \text{ else if } n \text{ odd then } f(3n+1) \text{ else } f(n/2)$$

a total function? Although so simple to state, this problem is still open, and so we cannot prove a tableau for the formula in the current state of mathematical knowledge.

This example suggests that arc weights greater than unity are trouble-makers, and so marked nets are an obvious class to study. Some other restrictions will be required (the register machine simulators are marked nets), but by putting further restrictions on either the nets or the formulae, there is hope of obtaining positive results.

References

[Acz77] P. Aczel, An introduction to inductive definitions. In: J. Barwise (ed.) *Handbook of Mathematical Logic* 739–782. North-Holland, Amsterdam (1977).

[Bak80] J. W. de Bakker, *Mathematical Theory of Program Correctness.* Prentice-Hall, Englewood Cliffs, NJ (1980).

[BaR72] J. W. de Bakker and W. De Roever, A calculus for recursive program schemes. *Proc. 1st Int. Coll. on Automata, Languages and Programming* 167–196. North-Holland, Amsterdam (1972).

[BKP84] H. Barringer, R. Kuiper and A. Pnueli, Now you may compose temporal logic specifications. *Proc. 16th ACM Symp. on Theory of Computing* 51–63 (1984).

[BHP82] M. Ben-Ari, J. Halpern and A. Pnueli, Deterministic propositional dynamic logic: finite models, complexity and completeness. *J. Computer and System Science* **25** 402–417 (1982).

[Ber86] G. Berthelot, Transformations and decompositions of nets. *Advances in Petri Nets 1986, Part I* 359–376. LNCS 254 (1986).

[Bes86] E. Best, Structure theory of Petri nets: the free choice hiatus. *Advances in Petri Nets 1986, Part I* 168–205. LNCS 254 (1986).

[BeV84] E. Best and K. Voss, Free choice systems have home states. *Acta Informatica* **21** 89–100 (1984).

[Bra87] J. C. Bradfield, *Verification of Concurrent Systems Represented as Petri Nets.* Dissertation submitted for the Diploma in Computer Science at the University of Cambridge (1987).

[Bra91] J. C. Bradfield, Proving temporal properties of Petri nets. To appear in *Advances in Petri Nets 1990.* LNCS (1991).

[BrS90] J. C. Bradfield and C. P. Stirling, Verifying temporal properties of processes. *Proc. CONCUR '90* 115–125. LNCS 458 (1990).

[BrS91] J. C. Bradfield and C. P. Stirling, Local model checking for infinite state spaces. To appear in *Theoret. Comput. Sci.* (1991).

[BCG89] M.C. Browne, E. M. Clarke and O. Grumberg, Reasoning about networks with many identical finite state processes. *Information and Computation* **81** 13–31 (1989).

[ClE81] E. M. Clarke and E. A. Emerson, Design and synthesis of synchronization skeletons using branching time temporal logic. LNCS 131

52–71 (1981).

[CES86] E. M. Clarke, E. A. Emerson and A. P. Sistla, Automatic verification of finite-state concurrent systems using temporal logic specifications. *ACM Trans. on Programming Languages and Systems* **8** 244–263 (1986).

[CPS89] R. Cleaveland, J. Parrow and B. Steffen, A semantics-based verification tool for finite state systems. *Proc. 9th IFIP Symp. on Protocol Specification, Testing and Verification.* North-Holland (1989).

[Com72] F. Commoner, Deadlocks in Petri nets. Report CA–7206–2311, Applied Data Inc. (1972).

[Coo78] S. A. Cook, Soundness and completeness of an axiom system for program verification. *SIAM J. Comp.* **7** (Feb. 1978).

[Dam90] M. F. Dam, Translating CTL* into the modal μ-calculus. Technical Report ECS–LFCS–90–123, LFCS, Dept. of Computer Science, University of Edinburgh, Edinburgh, U.K. (1990).

[EmL86] E. A. Emerson and C.-L. Lei, Efficient model checking in fragments of the propositional mu-calculus. *Proc. First IEEE Symp. on Logic in Computer Science* 267–278 (1986).

[Esp90] J. Esparza, Synthesis rules for Petri nets, and how they lead to new results. *Proc. CONCUR '90* 182–198. LNCS 458 (1990).

[FiL79] M. J. Fischer and R. E. Ladner, Propositional dynamic logic of regular programs. *J. Computer and System Science* **18** 194–211 (1979).

[Fit83] M. Fitting, *Proof Methods for Modal and Intuitionistic Logics* (Reidel, 1983).

[Hac72] M. Hack, Analysis of production schemata by Petri nets. TR–94, MIT–MAC (1972).

[HaR81] J. Y. Halpern and J. H. Reif, Propositional dynamic logic of deterministic, well-structured programs. *Proc. 22nd Symp. on Foundations of Computer Science* 322–334. Nashville, Tennessee (1981).

[HPS81] D. Harel, A. Pnueli and J. Stavi, Propositional dynamic logic of context-free programs. *Proc. 22nd Symp. on Foundations of Computer Science* 310–321. Nashville, Tennessee (1981).

[Flo67] R. Floyd, Assigning meanings to programs. In J. T. Schwartz, *Mathematical Aspects of Computer Science* 19–32. American Mathematical Society, Providence (1967).

[Hac76] M. Hack, The equality problem for vector addition systems is undecidable. *Theoret. Comput. Sci.* **2** 77–95 (1976).

[HoP79] J. Hopcroft and J.-J. Pansiot, On the reachability problem for 5-dimensional vector addition systems. *Theoret. Comput. Sci.* **8** 135–159 (1979).

[Hoa69] C. A. R. Hoare, An axiomatic basis for computer programming. *Comm. of the ACM* **12** 576–580 (1969).

[Hüt90] H. Hüttel, *SnS* can be modally characterized. *Theoret. Comput. Sci.* **74** 239–248 (1990).

[HüS91] H. Hüttel and C. P. Stirling, Actions speak louder than words: proving bisimilarity for context-free processes. Submitted for publication (1991).

[Koz83] D. Kozen, Results on the propositional mu-calculus. *Theoret. Comput. Sci.* **27** 333–354 (1983).

[KoP81] D. Kozen and R. Parikh, An elementary proof of the completeness of PDL. *Theoret. Comput. Sci.* **14** 113–118 (1981).

[KoP83] D. Kozen and R. Parikh, A decision procedure for the propositional mu-calculus. *Second Workshop on Logics of Programs* (1983).

[Lar90] K. Larsen, Proof systems for satisfiability in Hennessy–Milner logic with recursion. *Theoret. Comput. Sci.* **72** 265–288 (1990).

[Maz88] A. Mazurkiewicz, Compositional semantics of pure place/transition systems. *Advances in Petri Nets 1988* 307–330. LNCS 340.

[MaP69] Z. Manna and A. Pnueli, Formalization of properties of recursively defined functions. *Proc. ACM Symp. on Theory of Computing* 201–210 (1969).

[MaP83] Z. Manna and A. Pnueli, How to cook a temporal proof system for your pet language. *Proc. 10th ACM Symp. on Principles of Programming Languages* 141–154 (1983).

[MaP89] Z. Manna and A. Pnueli, The anchored version of the temporal framework. *Proc. Workshop on Linear Time, Branching Time and Partial Order in Logics for Concurrency.* LNCS 354 (1989).

[MeM88] J. Meseguer and U. Montanari, Petri nets are monoids. *LICS '88.* Computer Society Press, Washington (1988).

[Par70] D. M. R. Park, Fixpoint induction and proof of program semantics. *Machine Intelligence 5* 59–78. Edinburgh University Press, 1970.

[Pnu81] A. Pnueli, Temporal semantics of concurrent programs. *Theoret. Comput. Sci.* **27** 333–354 (1983).

[Pra76] V. Pratt, Semantical considerations of Floyd–Hoare logic. *Proc. 1th IEEE Symp. on Foundations of Computer Science* 109–121

(1976).

[Pra81] V. Pratt, A decidable mu-calculus. *Proc. 22nd Ann. ACM Symp. on Foundation of Computer Science* 421–427 (1981).

[Rei85] W. Reisig, *Petri Nets*. EATCS Monographs on Theoretical Computer Science. Springer–Verlag, Berlin–Heidelberg–New York (1985).

[Rei86] W. Reisig, Place/Transition Systems. *Advances in Petri Nets 1986, Part I* 117–141. LNCS 254 (1986).

[Sco70] D. S. Scott, Outline of a mathematical theory of computation. *Proc. 4th Annual Princeton Conf. on Information Sciences and Systems* 169–176. Princeton University, Princeton (1970).

[ScS71] D. S. Scott and C. Strachey, Toward a mathematical semantics for computer languages. *Proc. Symp. on Computers and Automata* 19–46. Polytechnic Institute of Brooklyn Press, New York (1971).

[Seg68] K. Segerberg, Decidability of S4.1. *Theoria* **34** 7–20 (1968).

[SiC86] A. P. Sistla and E. M. Clarke, Complexity of propositional temporal logics. *J. ACM* **32** 733–749 (1986).

[SoM89] Y. Souissi and G. Memmi, Composition of nets via a communication medium. *Proc. 10th Int. Conf. on Theory and Application of Petri Nets* 292–311 (1989).

[Sti87] C. P. Stirling, Modal logics for communicating systems. *Theoret. Comput. Sci.* **49** 311–347 (1987).

[Sti89] C. P. Stirling, Temporal logics for CCS. *Proc. Workshop on Linear Time, Branching Time and Partial Order in Logics for Concurrency* LNCS 354 (1989).

[Sti91] C. P. Stirling, Modal and temporal logics. In S. Abramsky, D. Gabbay and T. Maibaum (eds.) *Handbook of Logic in Computer Science*. Oxford University Press (1991).

[StW89] C. P. Stirling and D. J. Walker, Local model checking in the modal mu-calculus. *Proc. International Joint Conference on Theory and Practice of Software Development* 369–382. LNCS 351 (1989).

[StW90] C. P. Stirling and D. J. Walker, A general tableau technique for verifying temporal properties of concurrent programs. *Proc. Int. BCS–FACS Workshop on Semantics for Concurrency* Workshops in Computing (Springer–Verlag, Berlin, 1990) (1–15).

[Str81] R. Streett, Propositional dynamic logic of looping and converse. *Proc. 13th Symp. on Theory of Computing* 375–383. Milwaukee, Wisconsin (1981).

[StE89] R. S. Streett and E. A. Emerson, An automata theoretic decision procedure for the propositional mu-calculus. *Information and Computation* **81** 249–264 (1989).

[Wal70] W. J. Walkoe, Jr, Finite partially-ordered quantification. *J. Symbolic Logic* **35** 535–555 (1970).

[Win84] G. Winskel, A new definition of morphism on Petri nets. *STACS '84*. LNCS 166 (1984).

[Win85] G. Winskel, Categories of models for concurrency. *Seminar on Concurrency, Pittsburgh 1984*. LNCS 197 (1985).

[Win88] G. Winskel, A category of labelled Petri nets and a compositional proof system. *LICS '88*. Computer Society Press, Washington (1988).

[Win90] G. Winskel, On the compositional checking of validity. *Proc. CONCUR'90* 481–501. LNCS 458 (1990).

[Wol83] P. Wolper, Temporal logic can be more expressive. *Information and Control* **56** 72–99 (1983).

List of Notations

Index

This index gives the occurrences of all but minor definitions (in bold) and of major references to other topics.

Progress in Theoretical Computer Science

Editor
Ronald V. Book
Department of Mathematics
University of California
Santa Barbara, CA 93106

Editorial Board

Erwin Engeler
Mathematik
ETH Zentrum
CH-8092 Zurich, Switzerland

Robin Milner
Department of Computer Science
University of Edinburgh
Edinburgh EH9 3JZ, Scotland

Gérard Huet
INRIA
Domaine de Voluceau-Rocquencourt
B. P. 105
78150 Le Chesnay Cedex, France

Maurice Nivat
Université de Paris VII
2, place Jussieu
75251 Paris Cedex 05
France

Jean-Pierre Jouannaud
Laboratoire de Recherche
  en Informatique Bât. 490
Université de Paris-Sud
Centre d'Orsay
91405 Orsay Cedex, France

Martin Wirsing
Universität Passau
Fakultät für Mathematik
  und Informatik
Postfach 2540
D-8390 Passau, Germany

Progress in Theoretical Computer Science is a series that focuses on the theoretical aspects of computer science and on the logical and mathematical foundations of computer science, as well as the applications of computer theory. It addresses itself to research workers and graduate students in computer and information science departments and research laboratories, as well as to departments of mathematics and electrical engineering where an interest in computer theory is found.

The series publishes research monographs, graduate texts, and polished lectures from seminars and lecture series. We encourage preparation of manuscripts in some form of TeX for delivery in camera-ready copy, which leads to rapid publication, or in electronic form for interfacing with laser printers or typesetters.

Proposals should be sent directly to the Editor, any member of the Editorial Board, or to: Birkhäuser Boston, 675 Massachusetts Avenue, Cambridge, MA 02139.

NEW IN 1991:

1. Leo Bachmair, *Canonical Equational Proofs*
2. Howard Karloff, *Linear Programming*
3. Ker-I Ko, *Complexity Theory of Real Functions*
4. Guo-Qiang Zhang, *Logic of Domains*
5. Thomas Streicher, *Semantics of Type Theory: Correctness, Completeness and Independence Results*
6. Julian Charles Bradfield, *Verifying Temporal Properties of Systems*